MATH STUDY SKILLS WORKBOOK

FOURTH EDITION

Your Guide to Reducing Test Anxiety
and Improving Study Strategies

PAUL D. NOLTING, Ph.D.
Learning Specialist

BROOKS/COLE
CENGAGE Learning™

Australia • Brazil • Japan • Korea • Mexico • Singapore • Spain • United Kingdom • United States

BROOKS/COLE
CENGAGE Learning

Math Study Skills Workbook, Fourth Edition
Paul D. Nolting

Acquisition Editor: Marc Bove

Developmental Editor: Stefanie Beeck

Assistant Editor: Shaun Williams

Editorial Assistant: Zack Crockett

Marketing Manager: Gordon Lee

Marketing Coordinator: Shannon Myers

Marketing Communications Manager:
Darlene Macanan

Content Project Manager: PreMediaGlobal

Design Director: Rob Hugel

Senior Art Director: Vernon Boes

Print Buyer: Mary Beth Hennebury

Rights Acquisitions Specialist, Text: Dean
Dauphinais

Rights Acquisitions Specialist, Images: Dean
Dauphinais

Production Service: PreMediaGlobal

Photo Researcher: Martha Hall

Cover Designer: Denise Davidson

Cover Photo: Garry Black/Masterfile

Compositor: PreMediaGlobal

Photo Credits: p. 1, Lisa F. Young, 2010; p. 13,
Monkey Business Images, 2010; p. 31, Cristian
M., 2010; p. 51, Sean Prior, 2010; p. 67, Francesco
Ridolfi, 2010; p. 87, lightpoet, 2010; p. 107,
Laurence Gough, 2010

For product information and technology assistance, contact us at
Cengage Learning Customer & Sales Support, 1-800-354-9706

For permission to use material from this text or product,
submit all requests online at **www.cengage.com/permissions**
Further permissions questions can be emailed to
permissionrequest@cengage.com

Library of Congress Control Number: 2010936404

ISBN-13: 978-0-8400-5309-1

ISBN-10: 0-8400-5309-6

Brooks/Cole
20 Davis Drive
Belmont, CA 94002-3098
USA

Cengage Learning is a leading provider of customized learning solutions with
office locations around the globe, including Singapore, the United Kingdom,
Australia, Mexico, Brazil, and Japan. Locate your local office at
www.cengage.com/global

Cengage Learning products are represented in Canada by Nelson Education, Ltd.

To learn more about Brooks/Cole, visit **www.cengage.com/brookscole**

Purchase any of our products at your local college store or at our preferred
online store **www.cengagebrain.com**

Printed in the United States of America
3 4 5 6 7 16 15 14

CONTENTS

How to Improve Your Reading, Homework, and Study Techniques   67

How to Remember What You Have Learned   87

How to Improve Your Math Test-Taking Skills   107

PREFACE

Wouldn't it be nice if all we had to do was listen to a lecture on math and read the textbook in order to learn it? That would be paradise. However, most math courses take place on Earth, and students have to do much more than just take notes and read a textbook. They need a system of study skills that will help them understand and master mathematics.

Many students in all levels of math courses have difficulty learning math because it is one of the most difficult subjects in college. First, many students who have struggled with math before going to college continue to struggle when they take their first college math courses, even developmental math courses. Second, some students who have done well in the past begin to struggle when they take upper-level math courses such as college algebra or calculus. They made As and Bs in previous math courses, but all of a sudden they are failing upper-level math courses. These students probably lived off their intelligence until they took math courses that challenged them. Since they never had to study for math, they did not know where to begin to study. Students in all levels of math courses benefit from designing a system of study skills.

Does this sound far-fetched? Not really. I have watched calculus students read this workbook and improve their grades in order to be more competitive candidates for engineering programs. For example, I asked two calculus II students why they were taking my math study skills course. I thought they needed a 1-credit-hour course to allow them to graduate, but I was wrong. Both students made a C in calculus I and said that they needed to make an A or B in calculus II because they were engineering majors. In most cases, students who make Cs in calculus are not admitted to engineering schools. Both students made an A in my course and a B in

their calculus II course and went on to engineering school. Success for them!

Another success story belongs to a student who failed a beginning algebra course three times. Yes, three times. She came to my office ready to quit college altogether. I convinced her to focus on designing a system of study skills before taking the course one more time. She did. With her new system of study skills, she passed with a B and stayed in college. Success for her!

Earning better math grades doesn't have to be the only benefit from spending time developing a system for studying math. According to many students, they were able to take the math study skills and adapt them to their other courses. In fact, some students claimed it was easier than adapting general study skills to learning math. They reported that their other course grades also went up. What a good experience!

So, what kind of assistance do most students want? Students want tips and procedures they can use to help them improve their math grades. The math study suggestions, however, have to be based on research and be statistically proven to improve student learning and grades. *Math Study Skills Workbook* is based on *Winning at Math: Your Guide to Learning Mathematics Through Successful Study Skills* (2002), which is the only math study skills text that can boast of statistical evidence demonstrating an improvement in students' ability to learn math and make better grades. Learning and using these study skills techniques will improve your grades.

Math Study Skills Workbook is designed to supplement your math course or study skills course or to be used as part of a math anxiety workshop. The workbook is designed for independent work or to be supplemented by lectures. To gain the most benefit, the workbook needs to be completed by midterm.

v

This workbook is designed to enhance learning by teaching math learning skills in small chunks. After each section, you are required to recall the most important information by writing it down. The writing exercises are especially designed to help develop personalized learning techniques, such as the 10 steps to better test taking. Each chapter review is designed to reinforce your learning and to help you immediately select and use the best learning strategy from that chapter. *Note that the location of the answers to the first three questions in each of the first three Chapter Reviews are given immediately following the question.* Lecture Notes pages have been added, where possible, to allow you to include your notes in this workbook.

Dan's Take is a new addition to each chapter. Dan recently graduated from college, so his experiences are fresh and relevant. His input at the end of each chapter helps each reader relate to the information in a more personal way. Make sure you read each Dan's Take.

As you adapt and practice each study skill, work it into a larger system for studying math. Decide which study skills will help with each of the following tasks you must perform when learning math:

1. Taking and reworking lecture notes
2. Learning vocabulary
3. Completing homework in a way that helps you learn and master the math
4. Mastering and memorizing the concepts
5. Preparing for tests
6. Taking tests
7. Managing any anxiety, whether math or text anxiety

Remember, to reap the most benefit, you need to complete this workbook and be using your system of study skills by midterm. Practice and master! Then, maybe we can use your success story in our next edition of the workbook!

What You Need to Know to Study Math

CHAPTER

1

M athematics courses are not like other courses. Because they are different, they require different study procedures. Passing most of your other courses requires only that you read, understand, and recall the subject material. To pass math, however, an extra step is required: you must use the information you have learned to solve math problems correctly.

Learning general study skills can help you pass most of your courses, but special math study skills are needed to help you learn more and get better grades in math.

CHAPTER GOALS

What You Will Learn:

1.1 Why Learning Math Is Different from Learning Other Subjects

1.2 The Differences Between High School and College Math

1.3 Why Your First Math Test Is Very Important

1

DAN'S TAKE

In high school, I was 6 feet 4 inches tall and weighed all of 160 pounds. Being thin would not have been so bad, but I was the starting power forward for my school's basketball team and usually drew the biggest guy on the opposing team to guard. This left me at a disadvantage while playing basketball with guys who weighed 60 pounds more than me.

My first season, I spent considerably more time on my backside than I did on my feet. It wasn't out of the ordinary for me to look up from the floor and see opposing fans laughing at me. And who could blame them really? I was thoroughly outmatched by opponents twice as wide and three times as frightening as I was, and the sight of me falling might well have been ripped straight out of a Saturday morning cartoon. In short, it was embarrassing.

My initial reaction to this embarrassment was to let it affect both my behavior and my self-esteem. I accepted my shortcomings and instead of addressing them, I justified them as being out of my control. After all, I could not put on 30 pounds overnight.

Then I had an epiphany. While I was able to do just enough on the court to hold my position on the roster, it became increasingly clear that I had work to do. Instead of feeling sorry for myself, I needed to find solutions. Since I was not very big, I had to look for other ways to improve. I knew there were other skills to focus on, and there were specific actions I could take to improve them. I ran up and down the driveway to improve my lung strength. I spent hours shooting free throws in the gym before games. I took a basketball with me everywhere I went to improve my ball handling. If any of these things made me even slightly better, the hard work would be worth it to me.

And for the most part it worked. I never did put on the extra weight I wanted, but I did become slightly quicker, more elusive. When I got knocked down, I'd pick myself right back up again and make sure to counteract the embarrassment with a key rebound or a clutch free throw. In fact, my shortcomings became the motivation for me to succeed. I don't know that I could have thrived without them.

Math, not unlike basketball, requires daily practice. For some of us, math does not come easily, and that is okay. As you read my stories throughout this workbook, you will quickly learn that I'm no mathlete. But I've also learned to make do. The fact is, there are ways to improve our math success. The goal of this book is to help students focus on how they can improve. It teaches very specific solutions to very specific problems. Whether one is an "A" student or a "wants to be an A" student, there is always something one can improve on. We hope that this workbook will help us do just that.

1.1 Why Learning Math Is Different from Learning Other Subjects

In a math course, you must be able to do four things:

1. *Understand* the material.
2. *Process* the material.
3. *Apply* what you have learned to solve a problem correctly.
4. *Remember* what you have learned in order to learn new material.

Of these four tasks, applying what you have learned to solve a problem correctly is the hardest.

EXAMPLES

Political science courses require that you learn about politics and public service. But your instructor will not make you run for governor to pass the course. Psychology courses require you to understand the concepts of different psychology theories. But you will not have to help a patient overcome depression to pass the course. In math, however, you must be able to solve problems correctly to pass the course.

Sequential Learning Pattern

Another reason that learning math is different from learning other subjects is that it follows a *sequential learning pattern,* which simply means that the material learned on one day is used the next day and the next day, and so forth. This building-block approach to learning math is the reason it is difficult to catch up when you fall behind. All building blocks must be included to be successful in learning math. You can compare learning math to building a house. Like a house, which must be built foundation first, walls second, and roof last, math must be learned in a specific order. Just as you cannot build a house roof first, you cannot learn to solve complex problems without first learning to solve simple ones.

EXAMPLE

In a history class, if you study Chapters 1 and 2, do not understand Chapter 3, and end up studying and taking a test on Chapter 4, you *could* pass. Understanding Chapter 4 in history is not totally based on comprehending Chapters 1, 2, and 3. To succeed in math, however, each previous chapter has to be completely understood before you can continue to the next chapter.

Sequential learning also affects your ability to study for math tests. If you study Chapter 1 and understand it, study Chapter 2 and understand it, and study Chapter 3 and *do not* understand it, then when you study for a test on Chapter 4, you are not going to understand it either, and you probably will not do well on the test.

 REMEMBER

To learn the new math material for the test on Chapter 5, you must first go back and learn the material in Chapter 4. This means you will have to go back and learn Chapter 4 while learning Chapter 5. (The best of us can fall behind under these circumstances.) However, if you do not understand the material in Chapter 4, you will not understand the material in Chapter 5 either, and you will fail the test on Chapter 5. This is why the sequential learning of math concepts is so important.

The sequential learning pattern is also affected by

- your previous math course grade
- your math placement test scores
- the time elapsed since your last math course

Sequential learning is influenced by how much math knowledge you have at the beginning of your course. Students who have forgotten or never acquired the necessary skills from their previous math courses will have difficulty with their current math courses. If you do not remember what you learned in your last math course, you will have to relearn the math concepts from the previous course as well as the new material for the current course. In most other courses, such as the humanities, previous course knowledge is not required. However, in math you must remember what the last course taught you so that you are prepared for the current course. Measuring previous course knowledge will be explained in Chapter 2, "How to Discover Your Math Learning Strengths and Challenges."

Math placement scores also affect sequential learning. If you barely scored high enough to be placed in a math course, then you will have math learning gaps. Learning problems will occur when new math material is based on one of your learning gaps. The age of the placement test score also affects sequential learning. Placement test scores are designed to measure your *current* math knowledge and are to be used immediately.

Sequential learning is interrupted if math courses are taken irregularly. Math courses are designed to be taken one after another. By taking math courses each semester, without semester breaks between courses, you are less likely to forget the concepts required for the next course. Research has shown that the more time between math courses, the more likely it is that a student will fail the current math course.

Now that you understand the building-block nature of math, think about your math history.

- What were your previous math grades?
- How well did you do on the math placement test at your college?

- How long has it been since you took a math course?
- When you look at your math history, were there semesters when you did not take math?

These questions are important because if there was too much time between your different math courses, you may have forgotten important math concepts that you need in your current class. To use the building-block analogy, the blocks may not be as strong any more.

Now that you understand how learning math is a building experience, what should you do? *First,* don't get anxious. Stay calm. *Second,* if your college has a diagnostic math inventory in the tutoring center or math lab, take it to see what math concepts you have forgotten. Then ask your instructor where you can go to relearn these math concepts. *Third,* take the time to follow through. Many students just give up too easily, or they think they will catch up a little at a time. Don't think that and don't give up. The energy put into your class at the beginning of the semester will be more productive than energy put into class at the end of the semester when you try to learn everything during the last week before the final exam. *Fourth,* study and really learn the math; don't just practice mimicking it. *Finally,* when it is time to register for the next semester, register immediately so you will be able to get into the math class you need. Why do all this? Because math is sequential!

Math as a Foreign Language

Another helpful technique for studying math is to consider it a foreign language. Looking at math as a foreign language can improve your study procedures. If you do not practice a foreign language, what happens? You forget it. If you do not practice math, what happens? You are likely to forget it, too.

Students who excel in a foreign language must practice it at least every other day. The same study skills apply to math, because it is considered a foreign language. Like a foreign language, math has unfamiliar vocabulary words or terms that must be put in sentences called equations. Understanding and solving a math equation is similar to speaking and understanding a sentence in a foreign language.

> **EXAMPLE**
>
> Math sentences use symbols (which stand for spoken words), such as
>
> = (for which you *say* "equals")
>
> − (for which you *say* "less")
>
> *x* (for which you *say* "unknown")

Learning *how* to speak math as a language is the key to math success. Currently, most universities consider computer and statistics (a form of math) courses as foreign languages. Some universities now actually classify math as a foreign language.

Math as a Skill Subject

Math is a *skill subject,* which means you have to practice actively the skills involved to master it. Learning math is similar to learning to play a sport, learning to play a musical instrument, or learning auto mechanics skills. You can listen to and watch your coach or instructor all day, but unless you *practice* those skills yourself, you will not learn.

> **EXAMPLES**
>
> In basketball, the way to improve your free throws is to *see and understand* the correct shooting form and then *practice* the shots yourself. Practicing the shots improves your free-throw percentage. However, if you simply listen to your coach describe the correct form and see him demonstrate it, but you do not practice the correct form yourself, you will not increase your shooting percentage.
>
> Let's say that you want to learn how to play the piano. You are thrilled when you manage to convince the best piano player in the world to be your instructor. During your first session, you intently stare at your teacher's hands as she moves her fingers across the keys, and you take note of every movement she makes and every note she plays. Now it is your turn. Is it plausible that just by watching her play, you will now be able to perform just as well as she did? No. You obviously cannot play the piano correctly just by watching someone, even if that person is the greatest player in the world. The only way to learn to play is to practice.
>
> Learning math is no different from learning to play the piano. You can go to classes, listen to your professors, watch them demonstrate how to complete problems, and completely understand everything they are saying. But if you leave class and do not practice what you have learned, you simply will not learn math.

Many of your other courses can be learned by methods other than practicing. In social studies, for example, information can be learned by listening to your instructor, taking good notes, and participating in class discussions. Many students mistakenly believe that math can be learned the same way.

REMEMBER

Math is different. If you want to learn math, you must practice. Practice not only means doing your homework but also means spending the time it takes to understand the reasons for each step in each problem.

Math as a Speed Subject

Math is a *speed subject,* which means that, in most cases, it is taught faster than your other subjects. Math instructors have a certain amount of material that must be covered each semester. They have to finish certain chapters because the next math course is based on the information taught in their courses. In many cases a common math department final exam is given to make sure you know the material for the next course. Instructors are under pressure to make sure you are ready for the final exam because it demonstrates how prepared you are for the next level in math. This is different from, let's say, a sociology course, in which an instructor's omission of the last chapter will not cause students too many problems in the next sociology or social science course. So don't complain to the math instructor about the speed

of the course. Instead, improve your study skills so you can keep up!

Another way math is a speed subject is that most of the tests are timed, and many students think that they will run out of time. THIS CAUSES PANIC AND FEAR! This is different from most of your other courses, in which you generally have enough time to complete your tests, or in courses that have multiple-choice tests where you can start quickly bubbling the responses on the scantron sheet if you start running out of time. Students not only must understand how to do the math problems but also must learn the math well enough to complete the problems with enough speed to finish the test.

What makes me curious is, if students feel that they don't have enough time to complete a math test, why are most of them gone before the test is over? Sure, students who have learned the math thoroughly may complete the test early. That makes sense. Some students, however, leave either because they don't know the material or because they want to escape the anxious environment or work through the test carelessly.

So, since speed is an issue in learning math, what should you do? *First,* to use an analogy, start a daily "workout" program to "stay in shape" mathematically. Review, review, review as you learn new material. *Second,* practice doing problems within a time constraint. Give yourself practice tests.

REMEMBER

Passing math is your goal, regardless of your attitude.

SECTION 1.1   REVIEW

1. How does a sequential learning pattern affect math learning?

_____

_____

_____

Give two examples.

Example 1: _____

Example 2: _____

2. List two examples of how learning math is similar to learning a foreign language.

   Example 1: _____

   _____

   Example 2: _____

   _____

3. How is math similar to a skills subject?

   _____

4. Why are math study skills important at all levels of math? (See Preface.)

   _____

5. Math study skills can help you in other subjects, but general study skills usually cannot improve math learning. Why is this statement important? (See Preface.)

   _____

   _____

1.2   The Differences Between High School and College Math

College-level math courses can be two to three times as difficult as high school math courses. There are many reasons for the increased difficulty: course class-time allowance, the amount of material covered in a course, the length of a course, and the college grading system.

The first important difference between high school and college math courses is the length of time devoted to instruction each week. Most college math instruction, for the fall and spring semesters, has been cut to 3 hours per week; high school math instruction is provided 5 hours per week. Additionally, college courses cover twice the material in the same time frame as do high school courses. What is learned in one year of high school math is learned in one semester (4 months) of college math.

Simply put, in college math courses you are receiving less instructional time per week and covering twice the ground per course as you were in high school math courses. The responsibility for learning in college is the student's. As a result, most of your learning (and *practicing*) will have to occur outside of the college classroom.

College Summer Semester Versus Fall or Spring Semester and the Difference Between Night and Day

College math courses taught during summer semesters are more difficult than those taught during fall or spring. Furthermore, math taught during night courses is more difficult than math taught during day courses.

Students attending a 6-week summer math session must learn the information—and master the skills—two and a half times faster than as students attending regular, full-semester math sessions. Although you receive the same amount of instructional classroom time, there is less time to understand and *practice the skills* between class sessions.

Summer classes are usually 2 hours per day, 4 days per week (nighttime summer classes are 4 hours per night, 2 nights per week).

> **EXAMPLE**
>
> If you do not understand the lecture on Monday, then you have only Monday night to learn the material before progressing to more difficult material on Tuesday. During a night course, you have to learn and understand the material before the break; after the break, you will move on to the more difficult material—*that night.*

Because math is a sequential learning experience, where every building block must be understood before proceeding to the next block, you can quickly fall behind, and you may never catch up. In fact, some students become lost during the first half of a math lecture and never understand the rest of the lecture (this can happen during just one session of night class). This is called "kamikaze" math because most students do not survive summer courses.

If you *must* take a summer math course, take a 10- or 12-week *daytime* session so that you will have more time to process the material between classes.

Course Grading System

The grading system for math is different in college than in high school.

> **EXAMPLE**
>
> While in high school, if you make a D or a borderline D/F, the teacher more than likely will give you a D, and you may continue to the next course. However, in some college math courses, a D is not considered a passing grade, or if a D is made, the course will not count toward graduation.

College instructors are more likely to give a grade of N (no grade), W (withdrawal from class), or F if you barely know the material. This is because the instructor knows that you will be unable to pass the next course if you barely know the current one.

Most colleges require students to pass two college-level algebra courses to graduate. In most high schools, you may graduate by passing one to three math courses. In some college degree programs, you may even have to take four math courses and make no worse than Cs in all of them to graduate.

The grading systems for math courses are very precise compared with the grading systems for humanities courses.

> **EXAMPLE**
>
> In a math course, if you have a 79% average and you need 80% to get a B, you will get a C in the course. But if you make a 79% in English class, you may be able to talk to your instructor and do extra-credit work to earn a B.

Because math is an exact science and not as subjective as English, do not expect your math instructor to let you do extra work to earn a better grade. In college, there usually is no grade given for "daily work," as there often is in high school.

In fact, *your test scores may be the only grades that will count toward your final grade.* Therefore, you should not assume that you will be able to "make up" for a bad test score.

The Ordering of College Math Courses

College math courses should be taken *in order,* from the fall semester to the spring semester. If at all possible, avoid taking math courses from the spring to fall semesters. There is less time between the fall and spring semesters for you to forget the information. During the summer break, you are more likely to forget important concepts required for the next course and therefore experience greater difficulty.

SECTION 1.2 REVIEW

1. Compare the amount of college class time to high school class time.

_____

_____

2. How do college summer math courses differ from courses offered in the spring or fall?

_____

_____

3. How can the order of taking math courses affect your learning?

_____

_____

1.3   Why Your First Math Test Is Very Important

Making a high grade on the first major math test is more important than making a high grade on the first major test in other subjects. The first major math test is the easiest and, most often, the one for which the student is least prepared.

Students often feel that the first major math test is mainly a review and that they can make a B or a C without much study. These students are over-looking an excellent opportunity to make an A on the easiest major math test of the semester. (Do not forget that this test counts the same as the more difficult math tests that are still to come.)

The end of the semester, these students sometimes do not pass the math course or do not make an A because of their first major test grade. In other words, the first math test score was not high enough to pull up a low test score on one of the remaining major tests.

Studying hard for the first major math test and obtaining an A offers you several advantages:

- A high score on the first test can compensate for a low score on a more difficult fourth or fifth math test. All major tests have equal value in the final grade calculations.
- A high score on the first test can provide assurance that you have learned the basic math skills required to pass the course. This means you will not have to spend time relearning the misunderstood material covered on the first major test while learning new material for the next test.
- A high score on the first test can motivate you to do well. Improved motivation can cause you

to increase your math study time, which will allow you to master the material.

- A high score on the first test can improve your confidence for higher test scores. With more confidence, you are more likely to work harder on the difficult math homework assignments, which will increase your chances of doing well in the course.

What happens if, after all your studying, you make a low score on your first math test? You can still use this test experience to help you improve your next grade or to help determine if you are in the right math course. Your first math test, no matter what you make on it, can be used as a diagnostic test. Your teacher can review your test with you to see which type of math problems you got right and which ones you need to learn how to solve. It may be that you missed only a few concepts that caused the low score, and you can learn how to do these problems by getting help from the teacher, a learning resource center, or the math lab. However, you need to learn how to do these problems immediately so that you don't fall behind in the course. After meeting with your teacher, ask for the best way you can learn the concepts that are the bases of the missed problems and how to prepare for the next test. Even students who made Bs on the first math test can benefit by seeing the teacher.

If you made below 50 on your first math test, I suggest that it might be a good idea to drop back to a lower-level math course. Even though it might be beyond the first week of drop and add, many colleges and universities will let you drop back to a lower math class after the first major math test. Students who drop back get a good foundation in mathematics that helps them in their next math courses. On the other hand, I have *seen* students who insisted on staying in the math course but then repeated it several times before passing. Some of the students stopped taking the math course and dropped out of college. Dropping back to a lower-level math course and passing it is the smartest move. These students went on to become more successful in their math courses.

REMEMBER

It does not matter where you start as long as you graduate. Discuss this option with your teacher.

A Bad Math "Attitude"

Students' attitudes about learning math are different from their attitudes about learning their other subjects. Many students who have had bad experiences with math do not like math and have a bad attitude about learning it but can learn it. In fact, some students actually *hate* math, even though these same students have positive experiences with their other subjects and look forward to going to those classes.

Society as a whole reinforces students' negative attitudes about math. It has become socially acceptable not to do well in math. This negative attitude has become evident even in popular comic strips, such as *Peanuts*. The underlying message is that math should be feared and hated and that it is all right not to learn math.

This "popular" attitude toward math may reinforce your belief that it is all right to fail math. Such a belief is constantly being reinforced by others.

The bad math attitude is not a major problem, however. Many students who hate math pass it anyway, just as many students who hate history still pass it. The major problem concerning the bad math attitude is how you use this attitude. If a bad math attitude leads to poor class attendance, poor concentration, and poor study skills, then you have a bad math attitude *problem*.

SECTION 1.3 REVIEW

1. Why is your first math test supposed to be the easiest?

   _____

   _____

   _____

2. Give three reasons why your first math test is so important.

   First Reason: _____

   _____

   Second Reason: _____

   _____

   Third Reason: _____

   _____

3. What do you do if you fail your first test?

   Option 1: _____

   _____

Option 2: _____

_____

Option 3: _____

_____

4. List three ways a bad math attitude could affect your mathematics learning.

First Way:

_____

Second Way:

_____

Third Way:

_____

CHAPTER 1 REVIEW

1. In a math course, you must be able to do four things. You must _____ the material, _____ the material, _____ what you have learned to solve the problem, and _____ what you have learned in order to learn new material. (Answers on page 2)

2. Math requires _____, which means that one concept builds on the previous concept. (Answer on page 3)

3. Placement test scores are designed to measure your _____ math knowledge and are to be used _____. (Answers on page 3)

4. Like a foreign language, math has unfamiliar vocabulary words or terms that must be put in sentences called _____.

5. Learning how to speak math as a _____ _____ is one key to math success.

6. Keeping a _____ attitude about math will help you study more efficiently.

7. College math courses are _____ to _____ times as difficult as high school math courses.

8. Math courses are even more difficult than other courses because a grade of _____ or better is usually required to take the next course.

9. Math grading is very precise; in many cases, you cannot do _____ _____ to improve your grade.

10. If you fail your first math test, you need to make an appointment with your _____ to review your _____. Then you can decide if you should _____ to a lower-level math course.

What is the most important information you learned from this chapter?

_____

_____

_____

_____

How can you immediately use it?

_____

_____

_____

DAN'S TAKE REVIEW

1. How did the way Dan initially handled his situation on the basketball court harm his overall performance?

   _____

   _____

   _____

2. How was he able to overcome his problems, even those that he deemed "out of his control"?

   _____

   _____

   _____

3. How can he (or you) avoid similar problems in the future?

   _____

   _____

   _____

How to Discover Your Math Learning Strengths and Challenges

CHAPTER

2

Students can improve their math grades when they take the time to understand their strengths and challenges in learning math. Then, if they enhance their strengths and find solutions to their challenges in learning math, they will see more success.

Just as a mechanic conducts a diagnostic test on a car before repairing it, you need to analyze your own learning in order to repair it and get it running smoothly. No one wants the mechanic to spend time working on something that doesn't need to be fixed. That doesn't make sense. Likewise, you want to work on the challenges you have when learning math, not on what you already do right when studying. This chapter guides you through a "learning" diagnostic so that you can improve some of your learning strategies for math and keep using the ones that already work.

Areas of math strengths and challenges include math knowledge, level of test anxiety, study skills, study attitudes, motivation, and test-taking skills. Before we start identifying your math strengths and challenges, you need to understand what contributes to math academic success.

DAN'S TAKE

Because I majored in journalism, I was required to take only one math course in order to graduate. I decided to put it off until the first semester of my senior year. I had not taken a math course since my junior year of high school, but I figured I could take the easiest math class in the course listings and cruise until graduation.

Wow, was I wrong.

Little did I know that I had to pass a math placement test, even though it had been more than 4 years since I had looked at math textbook. Needless to say, I bombed. I was forced to take a remedial class before being allowed to take the course I needed to graduate.

What's more, the class I was placed in was full of freshman. Imagine my embarrassment being the only senior in a class of 18-year-olds, most of them assuming I was a fledgling college student myself. The whole semester I had to answer questions such as, Have you decided on a major yet? or How crazy was prom? For the duration of the class I contemplated wearing a T-shirt that said, "I'm a senior. Leave me alone."

That said, I fully realized the fault began and ended with me—or, more specifically, my decision-making process. My first mistake was waiting so long to take my math credit. I could have easily completed the course my freshman year when the information was still fresh in my head. The next mistake was that I assumed that I knew the math. I never measured my math knowledge, and I went into the placement test in total ignorance.

Then came the salt in the wound. After working in the remedial class for a couple of weeks, the math began to come back to me. It was easy—basic even. It became very clear that I could have passed the placement test with just a little preparation. There were numerous on-line programs I could have used, and there were books I could have checked out from the library, but instead I went into the test completely unprepared.

Lucky for me, I still had two full semesters before I graduated and, therefore, remained able to complete both the remedial class and the designated course required for my major. Otherwise, my lack of preparation might have cost me sizable amounts of both time and money.

The lesson to be learned here is that math is not a subject that should be put off. It is best to head into your new math course with the prerequisite knowledge freshly stored in your mind. Despite what you might think, college is not only finite, but it also moves by in a blink of the eye. By measuring what you know beforehand, you can avoid my situation and not waste your time and resources on classes you could otherwise avoid.

2.1   Variables Contributing to Student Academic Achievement

Dr. Benjamin Bloom, a famous researcher in the field of educational learning, discovered that your IQ (intelligence) and your cognitive entry skills account for 50% of your course grade. See Figure 1.

Quality of instruction represents 25% of your course grade, while affective student characteristics reflect the remaining 25% of your grade.

- *Intelligence* may be considered, for our purpose, as how fast you can learn or relearn math concepts.
- *Cognitive entry skills* refer to how much math you already know the first day of your new class. For example, previous math knowledge can be based on previous math grades, the length of time since the last math course, the type of math

courses completed (algebra I, algebra II, intermediate algebra), and placement test scores.

- *Quality of instruction* is concerned with the effectiveness of math instructors when presenting material to students in the classroom and math lab. This effectiveness depends on the course textbook, curriculum, teaching style, extra teaching aids (videos, audiocassettes), and other assistance. Quality of instruction may also include math lab resources, quality of tutoring based on tutor training, and Supplemental Instruction.
- *Affective student characteristics* are characteristics you possess that affect your course grades—excluding how much math you knew

FIGURE 1 Variables Contributing to Student Academic Achievement

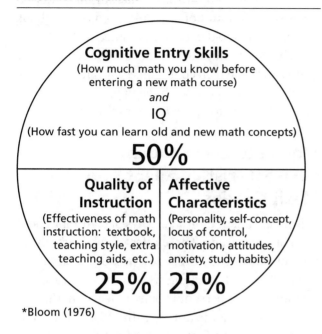

Cognitive Entry Skills
(How much math you know before entering a new math course)
and
IQ
(How fast you can learn old and new math concepts)
50%

Quality of Instruction
(Effectiveness of math instruction: textbook, teaching style, extra teaching aids, etc.)
25%

Affective Characteristics
(Personality, self-concept, locus of control, motivation, attitudes, anxiety, study habits)
25%

*Bloom (1976)

before entering your math course. Some of these affective characteristics are anxiety, study skills, study attitudes, self-concept, motivation, and test-taking skills.

The following sections help you examine your own variables for success in math.

Cognitive Entry Skills

Poor math knowledge can cause low grades. A student placed in a math course that requires a more extensive math background than the student possesses will probably fail that course. Without the correct math background, you will fall behind and never catch up.

Placement Tests and Previous Course Grades

The math you need to know to enroll in a particular math course can be measured by a placement test (ACT, SAT) or by the grade received in the prerequisite math course. Some students are incorrectly placed in math courses on the basis of placement tests.

If, by the second class meeting, everything looks like Greek and you do not understand what is being explained, move to a lower-level course. In the lower-level math course you will have a better chance

of understanding the material and of passing the course. A false evaluation of math ability and knowledge can only lead to frustration, anxiety, and failure.

Requests by Students for Higher Placement

Some students try to encourage their instructors to move them to higher-level math courses because they believe their placement scores were wrong. Many students try to avoid noncredit math courses, while other students do not want to repeat courses that they have previously failed. These moves can also lead to failure.

Some older students imagine that their math skills are just as good as when they completed their last math course, which was 5 to 10 years ago. If they have not been practicing their math skills, they are just fooling themselves. Still other students believe they do not need the math skills obtained in a prerequisite math course to pass the next course. This is also incorrect thinking. Research indicates that a student who was placed correctly in the prerequisite math course and who subsequently failed it will not pass the next math course.

What My Research Shows

I have conducted research on thousands of students who have either convinced their instructors to place them in higher-level math courses or have placed themselves in higher-level math courses. The results? These students failed their math courses many times before realizing they did not possess the prerequisite math knowledge needed to pass the course. Students who, without good reason, talk their instructors into moving them up a course level are setting themselves up to fail.

To be successful in a math course, you must have the appropriate math knowledge. If you think you may have difficulty passing a higher-level math course, you probably do not have an adequate math background. Even if you do pass the math course with a D or a C, research indicates that you will most likely fail the next higher math course.

It is better to be conservative and pass a lower-level math course with an A or a B instead of making a C or a D in a higher-level math course and failing the next course at a higher level.

This is evident when many students repeat a higher-level math course up to five times before repeating the lower-level math course that they barely passed. After repeating the lower-level math course with an A or a B, these students passed the higher-level math course.

How the Quality of Math Instruction Affects Your Grades

Quality of instruction accounts for 25% of a student's grade. Quality of instruction includes such things as classroom atmosphere, instructor's teaching style, lab instruction, and textbook content and format. All these "quality" factors can affect a person's ability to learn in the math classroom.

Interestingly enough, probably the most important quality variable is the compatibility of an instructor's teaching style with your learning style. You need to discover your learning style and compare it with the instructional style. Noncompatibility can be best solved by finding an instructor who better matches your learning style. However, if you cannot find an instructor to match your learning style, improving your math study skills and using the math lab or learning resource center (LRC) can compensate for most of the mismatch.

Use of the math lab or LRC can dramatically improve the quality of instruction. With today's new technologies, you are able to select your best learning aids. These learning aids could be DVDs, CD-ROMs, computer programs, math study skills computer programs, math texts, and Websites.

The quality of tutors is also a major factor in the math lab or LRC. A low student-to-tutor ratio and trained tutors are essential for good tutorial instruction. Otherwise, the result is just a math study hall with a few helpers.

The math textbook should be up to date with good examples and a solutions manual. This increases the amount of learning from the textbook compared with older, poorly designed texts.

The curriculum design affects the sequence of math courses, which could cause learning problems.

Some math courses have gaps between them that cause learning problems for all students.

How Affective Student Characteristics Affect Your Math Grades

Affective student characteristics account for about 25% of your grade. These affective characteristics include math study skills, test anxiety, motivation, locus of control, learning style, and other variables that affect your personal ability to learn math.

Most students do not have this 25% of the grade in their favor. In fact, most students have never been taught any study skills, let alone how to study math specifically. Students also do not know their best learning styles, which means they may study ineffectively by using their least effective learning styles. However, it is not your fault that you were not taught math study skills or made aware of your learning styles.

By working to improve your affective characteristics, you will reap the benefits by learning math more effectively and receiving higher grades. Thousands of students have improved their math affective characteristics and, thereby, their grades by using this workbook.

This section identified key factors that can determine a student's success in a math course. The next section focuses on your personal "learning diagnostic" for learning math.

SECTION 2.1 REVIEW

1. List and explain the three variables that contribute to your academic success.

Variable 1: _____

_____

_____

_____

Variable 2: _____

_____

_____

_____

Variable 3: _____

_____

_____

_____

2. How does your placement test or previous math course grade affect your math learning?

_____

_____

_____

_____

3. How does the quality of math instructors affect your grade?

_____

_____

_____

_____

4. How do your affective characteristics affect your grade?

_____

_____

_____

_____

2.2   Your Learning Diagnostic for Math Class

Research has shown that matching a student's best learning style with the instructional style improves learning. Research has also shown that students who understand their learning styles can improve their learning effectiveness. A learning disadvantage will occur for students who do not know or do not understand their learning styles. Students should talk to their instructors or counselors about taking one or more learning style inventories.

Measuring Previous Math Knowledge

Now it's time to measure your previous math knowledge. Your previous math knowledge is an excellent predictor of future math success.

Placement Test Scores

Visit your counselor or instructor to discuss your math placement test score. Ask your counselor or instructor how your placement score relates to the scores for the math courses below and above your course.

Find out if your score barely placed you in your course, if your score was in the middle, or if your score almost put you into the next course. If your placement score barely put you into your current math course, you will have difficulty learning the math required in that course.

A barely passing placement score means that you are missing some of the math knowledge you require to be successful, and you will really struggle to pass the course.

A middle placement score indicates that you probably have most of the math knowledge required for the course; however, you still may have difficulty passing the course. A high placement score means you have most of the math knowledge needed to pass and may have a good chance of passing the course. (Still, if you have poor math study skills and high test anxiety, you may not pass the course.)

Students who have low placement test scores need to improve their math knowledge as soon as possible. Students with middle placement scores should also consider improving their math knowledge.

Previous Grades

The best predictor of math success is your previous math grade (if taken within 1 year of your current course).

If you are taking your second math course, use your previous math course grade to determine your math knowledge. Your previous course grade is the best predictor of success in your next math course. (However, this is true only if your last math course was taken within the last year.)

A student who made an A or a B in the previous math course has a good chance of passing the next math course. However, a student who made a C in the previous math course has a poor chance of passing the next math course. A student who made less than a C or withdrew from the previous math course has virtually no chance of passing the next math course.

The Amount of Time Since Your Last Math Course

Another indicator of math success can be the length of time since your last math course. The longer the time since your last math course, the less likely you will be to pass the current course. The exception to this rule is if you were practicing math while you were not taking math courses. Math is very easily forgotten if not practiced.

How long has it been since your last math course? If it has been two or more years since your last math class, your math may be similar to someone who has a low placement score or made a C or lower in the last math class. You can improve your math knowledge by following the same suggestions made for those students and applying the information in the next paragraph.

What level of math knowledge preparation do you possess right now for this course? If you have poor or good previous math knowledge, you need to build up those math concepts and skills as soon as possible. Go to your instructor and discuss how to build up your skills. Some suggestions are to review the chapters of the previous math text or ask your instructor about some Websites that help students review math. Students with poor math knowledge have about 3 to 4 weeks to improve their math before it can become a major problem. Don't wait. Start now to review your math skills in order to improve your mathematics learning and grades. Additional information is presented in the section "How to Improve Your Math Knowledge."

Taking Stock of Your Learning Style

After examining your level of math knowledge and making any necessary accommodations, it is important to examine how you learn math best. There are many different learning styles, and no one style is necessarily better than another. Savvy students learn how to use their best learning styles and then learn how to use other learning style strategies, too. The first learning inventory you will take is a *Math Learning Modality Inventory*. It will measure how you take information in—visually, auditorily, or kinesthetically.

Name: _____ Date: _____

LEARNING MODALITY INVENTORY FOR MATH STUDENTS

The following survey can help you discover how you best learn math. Answer the questions based on your personal learning characteristics. There are no right or wrong answers. The more you answer truthfully, the more you will be able to use the results to improve your study of math. A "1" means that the statement is hardly like you. A "4" means that the statement is really like you. Then, if you think the statement is somewhere in between, decide if it is a "2" or a "3."

Questions	Least Like Me		Most Like Me	
	1	2	3	4
1. Reading a math problem out loud helps me learn better when I am studying.	1	2	3	4
2. I learn math better if I can talk about it.	1	2	3	4
3. I select certain problems and memorize what they look like so I can use them to help me remember on a math test.	1	2	3	4
4. Making things with my hands helps me learn better.	1	2	3	4
5. Drawing a picture of a word problem helps me understand how to do it on a test.	1	2	3	4
6. Math makes more sense when I see it worked out on the board.	1	2	3	4
7. Moving around while studying helps me concentrate and learn more.	1	2	3	4
8. I understand written instructions better than ones told to me.	1	2	3	4
9. I memorize what a problem looks like so I can remember it better on a test or quiz.	1	2	3	4
10. I repeat the steps of a problem out loud or to myself in order to remember what I am supposed to do.	1	2	3	4
11. Watching someone complete a math problem helps me understand more than listening to someone tell me how to do it.	1	2	3	4
12. Talking about a math problem while learning in class helps me understand it better.	1	2	3	4
13. I learn math better when I watch someone do it.	1	2	3	4
14. When I take a test, I read the problems to myself softly.	1	2	3	4
15. When I solve a math problem on a test, I picture my notes in my head to help me remember how to solve it.	1	2	3	4
16. I enjoy making things with my hands for a hobby.	1	2	3	4
17. Math makes more sense to me when someone talks about it while doing it on the board rather than just doing it on the board.	1	2	3	4
18. Explaining a math problem to someone else helps me learn better when I am studying.	1	2	3	4
19. Looking at a picture from my notes or math book helps me understand a math problem.	1	2	3	4
20. Making study aids with my hands helps me learn better.	1	2	3	4
21. I understand instructions better when someone tells me what they are.	1	2	3	4
22. I memorize sentences or words I can say to myself to help me remember how to do problems on a test.	1	2	3	4
23. Pictures and charts help me see how all the parts of a word problem work together.	1	2	3	4
24. I enjoy putting things together.	1	2	3	4
25. When I solve a problem on a math test, I talk my way through it in my head or softly to myself.	1	2	3	4

Scoring Your Results

Step 1: Fill in each answer score next to the appropriate question number. Add the column totals. Divide column totals A and B by 2. Those numbers will be your final column totals. Leave column C total as is.

Column A	Column B	Column C
1. _____	3. _____	4. _____
2. _____	5. _____	7. _____
10. _____	6. _____	16. _____
12. _____	8. _____	20. _____
14. _____	9. _____	24. _____
17. _____	11. _____	
18. _____	13. _____	
21. _____	15. _____	
22. _____	19. _____	
25. _____	23. _____	
A Total _____ /2 = _____ Column Total	B Total _____ /2 = _____ Column Total	C Total _____ (Do not divide.)

Step 2: Fill in the number of squares to represent each column total. Any total greater than 12 indicates that modality style as a strength when you learn math. You can be strong in more than one modality. If none of the totals equals 12 squares, your highest score is your strongest modality. If you have a tie, pick the one that comes to mind as your strongest.

	Least Like Me																	Most Like Me	
Modality	1			5					10					15					20
A = Auditory																			
B = Visual																			
C = Kinesthetic																			

This Learning Modality Inventory was developed by Kimberly Nolting, ©.

The Math Learning Modality Inventory

The Math Learning Modality Inventory measures how you best input new information when you are learning math. Some students understand new math better when they can see it worked out (*visual*), while others understand better when the math is explained using language (*auditory*). Still other students understand new math best when they have models or other tools that represent a math problem. They can use their hands to help them learn (*kinesthetic*). The word *modality* describes these different senses.

No one modality is necessarily better than the others. It is part of students' learning personalities. What is important, however, is to know your preferred learning modality and to design study strategies that use it. Even better, after learning how to use your preferred learning modality, learn how to use the other modalities when studying and learning in class, too. Why? Sometimes your instructor will teach in a way that is not your strongest way to learn. It is up to you to learn new strategies in that situation.

Cognitive Learning Styles

Cognitive learning styles describe how you process information, once you have heard, seen, or felt the information. Each individual has a unique system of doing this, but there are several general styles of what is called *cognitive processing*. Bernice McCarthy (1981) named four of these styles based on different ways to process information and make meaning from it: innovative learning, analytic learning, common sense learning, and dynamic learning.

First read through all the styles. Then read each style again, checking off the characteristics with which you identify. What do these cognitive styles mean when it comes to learning math? First, each of these types of learners can learn math. Second, analytic and commonsense learners tend to learn math using their natural ways to learn. Third, math teachers tend to be analytic and/or commonsense learners and teachers. Fourth, dynamic and innovative learners may need to learn how to think more analytically and use common sense, since it may not be their preferred way to learn math. They may grow impatient with all the details or feel that math is just not their subject when it really can be. Dynamic and innovative learners can be good mathematicians. They just may need to be more conscious of designing study strategies for math.

Chapter 6 suggests different study skills that will help you learn math through your preferred learning modality and style.

REMEMBER

It is important to recognize that students who do not have learning styles that easily lend themselves to learning math can still learn math. Good study skills and awareness on the part of the math instructor and student can compensate for a mismatch of modality and learning styles. Working together to find strategies to help improve the teaching and learning is most productive for everyone.

WHY? *Innovative learners* are imaginative thinkers with lots of ideas. They learn by thinking, listening, and sharing ideas. They want to make a personal connection to what they are learning, to the class, or to the instructors. They learn by bringing their own personal experiences into the process. They want to know Why? when they are learning. They solve problems by personally relating to them and using their feelings.

Innovative learners

- work toward self-improvement
- tap into feelings to interpret what is going on
- listen and reflect
- like subjective tests
- strive for peace and harmony with people
- have concern for others
- listen and share
- have a favorite question: Why?
- are interested in people and their feelings
- personalize what they learn
- have lots of ideas

They like careers and majors in fields such as

- counseling
- humanities
- social work

- education (in order to help others)
- English
- communications
- human resources
- other careers interacting with/or helping people

WHAT? *Analytic learners* put details together and use facts to understand ideas. They are more interested in ideas, facts, and theories than in people. They are uncomfortable with subjective topics and prefer logic. They learn best in traditional lecture classrooms when the instructor is the expert and provides the information. They want to know What? when they are learning. They solve problems by gathering information and using logic.

Analytic learners

- use facts to interpret what is going on
- like lectures by experts
- usually prefer studying alone, and not in groups
- prefer objective tests
- are interested in ideas and facts more than in people
- enjoy research
- must have as much information as possible to make decisions
- have a favorite question: What?
- form theories from information they collect
- want to know what the experts think
- like problem solving

They like careers and majors in fields such as

- mathematics
- education (in order to teach knowledge)
- sciences
- engineering or computer science
- accounting
- other careers that use logical reasoning

HOW? *Commonsense learners* want to know how they can immediately use what they are learning. They use facts to learn about concepts and want to immediately use the information and see immediate application. (Analytic learners are satisfied with just coming up with the theories and concepts.) Commonsense learners want to get right to the point when they are learning or studying. They want to know How does this work? They solve problems by knowing how something works.

Common sense learners

- believe things should make common sense
- test theories with applications
- learn in math and science labs

- want to know how something applies immediately
- do not tolerate fuzzy ideas
- have a "lightbulb" go on when applying new ideas
- are skills oriented and like tinkering with things
- have a favorite question: How?
- want to get right to the point
- like objective tests
- like classes that are practical

They like careers and majors in fields such as

- applied sciences
- engineering
- allied health sciences
- technologies
- computer science
- other fields in which they can take ideas and make practical applications

WHAT IF? *Dynamic learners* want to know what new possibilities exist for the concepts they are learning. They like learning to be more of a self-discovery process than a "listen to a lecture and discuss" approach. They like new things and can adapt to change. They learn by trial and error. They solve problems by looking at hidden possibilities. Their favorite question is What if?

Dynamic learners

- adapt and even like change
- learn by trial and error
- learn best by independent self-instruction
- test new ideas through experiences
- get bored with routine in learning
- need to know what else can be done with new information
- are at ease with people
- have a favorite question: What can this become?
- have a favorite question: What if?
- like self-discovery learning
- often reach correct answers intuitively (using the "gut")
- are risk takers

They like careers and majors in fields such as

- business
- administration (if innovation is allowed)
- marketing
- education (to help students learn to reach their goals)
- sales occupations
- roles or jobs in which they can get actions started and energize people

SECTION 2.2   REVIEW

1. My learning modality is _____ because _____

   _____

2. My cognitive learning style is _____ because _____

   _____

2.3   Measuring Your Math Study Skills

Students have learning strengths and challenges that affect their math learning and course grades. They must identify and understand these strengths and challenges to know which strengths support them and which challenges to improve upon. After reading this section, you will be asked to complete the Math Learning Profile Sheet (Figure 2).

> **EXAMPLE**
>
> Having good math study skills is a *positive* math learning characteristic, whereas having high test anxiety is a *negative* math learning characteristic.

Math study skills comprise the main part of the affective characteristics that represent about 25% of your math grade. You can determine your math study skills expertise by taking the free Math Study Skills Evaluation at www.academicsuccess.com, or you can do a self-determination of your math study skills. If you take the online Math Study Skills Evaluation, the results will indicate your score. If you score below 70, you really need to focus on developing a set of study skills. A score between 70 and 80 means that you have average study skills, and a score above 90 means that you have good study skills. A self-determination can be made by asking yourself if you have ever been taught math study skills. If you have, then you probably have either good or excellent math study skills. If you have not, then you most likely need to develop good math study skills.

Scoring low on the Math Study Skills Evaluation is not the end of the world. In fact, it might indicate that your problems with learning math may be due to your study habits instead of your math intelligence.

A low score on the Math Study Skills Evaluation means that most of your learning problems result from the fact that you have not been taught math study skills. This workbook can teach you math study skills. If you previously have gotten poor math grades, they were probably due to a lack of math study skills, and this was not your fault.

You can improve your math study skills and become a successful math student. If you have good math study skills, you can still improve them to make As in math. If you have excellent math study skills, you can still learn some additional math study skills that can help you in future courses.

Previous Study Skills Courses or Training

General study skills training that does not focus on math may help some students in time management, reading techniques, learning styles, and overcoming procrastination. Even though math requires specific study techniques, you can learn many other helpful general study skills and personal habits that will help you in math classes. If you have had no study skills training, then this workbook will help you in your math and other courses. Students with no past study skills training should also take advantage of any other opportunities to learn about college success skills in addition to learning from this workbook. Contact

FIGURE 2   Math Learning Profile Sheet

After completing this profile, use your answers to select which areas you would like to improve first. If you have a serious problem with math or test anxiety, get help in this area immediately. Circle the X that best describes your belief.

	Strongly Disagree	Disagree	Agree	Strongly Agree
1. I have good math study skills.	X	X	X	X
2. Math tests don't make me anxious.	X	X	X	X
3. I have a strong basic knowledge of math to pass this class.	X	X	X	X
4. I have a positive attitude about learning math.	X	X	X	X
5. It has been less than a year since I took my last math class.	X	X	X	X
6. I have taken a math study skills course before this class.	X	X	X	X
7. I have good reading skills to be successful in this class.	X	X	X	X
8. I know if I improve my study habits, I can improve my math grades.	X	X	X	X

This informal survey is designed to help you decide what kind of shape you are in to do well in your course.

Which areas are strong and will help you do well in a math class?

_____

_____

Which areas keep you from doing the best you can in a math class? How can you change them?

_____

_____

your learning centers, instructors, or counselors about study skills workshops and when they will be offered.

Previous or Current Required Reading Courses

College reading skills are also needed for success in math courses. This is especially true when you are trying to solve story or word problems.

At community colleges and some universities, there are three levels of reading courses. Usually, the first two levels are noncredit reading courses, and the third level is for college credit.

If you are enrolled in the lowest reading level, you will have difficulty reading and understanding the math text. Students who are enrolled in the second level may also have difficulty reading the text. Students who are not required to take a reading class or who are reading at grade level may still have some difficulty reading the math text.

Students not at college level reading can pass math courses; however, if you need help in reading, get that help as soon as possible. Or, many learning or study skills centers help students with reading skills for specific courses. Work with a learning specialist to design a system for reading your math textbook. If you do not know your reading level, ask to take a reading test. If in doubt about your reading level, take a reading class.

REMEMBER

Math texts are not written like English or history texts. Even students with college-level reading skills may experience difficulty understanding their math text.

2.4 Reflecting on Your Affective Attributes

Determining Your Math Attitude

Math attitude can play a major part in your success in math courses. Some students who have a poor attitude toward math may not attend class as often as they should or may procrastinate in doing their homework. Other students avoid math until they have to take it. What is your math attitude—good, neutral, or negative? Remember, students with negative math attitudes can pass math. It depends on what you do with your attitude that counts. Don't let your negative attitude cause problems in your math course. Also, if you have a good math attitude, don't forget to study. A good math attitude can only get you so far!

Measuring Your Math Anxiety Level

It is also important to assess any math anxiety you might have, so that you can start managing it immediately. If you took the online Math Study Skills Evaluation, your score should indicate if you "seldom," "often," or "almost always" become anxious and forget important concepts during a test.

If you have medium or high math test anxiety, it can be reduced. This problem can be fixed if you follow the suggestions in the next chapter. This also means that some of your problems in learning math (if you have any) are due to anxiety instead of any intellectual inability to learn math.

If you have high text anxiety, then you will probably need to use the *How to Reduce Test Anxiety* CD. You can order this CD by going to www.academicsuccess.com.

Your Locus of Control

Locus of control is a concept that describes how much control you feel you have over both your life and your course grades. Some students feel they have a lot of control over their lives and learning, while other students feel they have very little control over their lives and grades. For example, when a student with external locus of control does poorly on a test, the blame is put on the instructor, when in reality the student did not study well enough. On the other hand, the student with internal locus of control admits that studying more and paying more attention in class would have helped. Students can improve their internal locus of control and be successful in mathematics at the same time.

Evaluating Your Findings

Now that you understand some of the general characteristics that help students be successful, complete the informal Math Learning Profile Sheet (Figure 2). When you look at your profile, think of it as similar to going to a fitness center for the first time. A professional trainer asks questions, runs you through equipment, analyzes your strengths and challenges, and then designs a program for you. You are doing the same thing here.

SECTION 2.4 REVIEW

1. List your strengths for learning math.

_____

_____

_____

_____

_____

2. List three areas you will start improving right now.

    Area 1: _____

    _____

    Area 2: _____

    _____

    Area 3: _____

    _____

3. How will you improve these areas?

    Area 1: _____

    _____

    Area 2: _____

    _____

    Area 3: _____

    _____

2.5   How to Improve Your Math Knowledge

Instructors always operate on the premise that you finished your previous math course just last week; they do not wait for you to catch up on current material. It does not matter if your previous math course was a month ago or 5 years ago. Instructors expect you to know the previous course material—period.

Review Your Previous Math Course Material and Tests

There are several ways to improve your math knowledge. Review your previous math course material before attending your present math course. Look closely at your final exam to determine your weak areas. Work on your weak areas as soon as possible so they can become building blocks (instead of stumbling blocks) for your current course.

If it has been some time since your last math course, visit the math lab or learning resource center to locate review material. Ask the instructor if there are any computer programs that will assess your math skills to determine your strengths and challenges for your course. Review math videotapes on the math course immediately below your level. Also review any computer software designed for the previous math course.

Another way to enhance learning is to review the previous math course text by taking all the chapter review tests. If you score above 80% on one chapter review test, move on to the next chapter. A score below 80% means you need to work on that chapter before moving on to the next chapter. Get a tutor to help you with those chapters if you have trouble. Make sure you review all the chapters required in the previous course as soon as possible. If you wait more than 2 weeks to conclude this exercise, it may be too late to catch up (while learning new material at the same time).

Employ a Tutor

One last way to improve your cognitive entry skills is to employ a private tutor. If you have a history of not doing well in math courses, you may need to start tutorial sessions *the same week class begins.* This will give the tutor a better chance of helping you regain those old math skills.

You still need to work hard to relearn old math skills while continuing to learn the new material. If you wait 4 to 5 weeks to employ a tutor, it will probably be too late to catch up and do well or even pass the course.

REMEMBER

Tutorial sessions work best when the sessions begin during the first 2 weeks of a math course.

Schedule Math Courses Back to Back

Another way to maintain appropriate math knowledge is to take your math courses back to back. It is better to take math courses every semester—even if you do not like math—so that you can maintain sequential (linear) learning.

I have known students who have made Bs or Cs in a math class and who then waited 6 months to a year to take the next math course. Inevitably, many failed. These students did not complete any preparatory math work before the math course and were lost after the second chapter. This is similar to having one semester of Spanish, not speaking it for a year, then visiting Spain and not understanding what is being said.

The only exception to taking math courses back to back is taking a 6-week "kamikaze" math course (an ultracondensed version of a regular course), which should be avoided.

If you are one of the unfortunate many who are currently failing a math course, you need to ask yourself, Am I currently learning any math or just becoming more confused? If you are learning some math, stay in the course. If you are getting more confused, withdraw from the course. Improve your

math knowledge prior to enrolling in a math course during the next semester.

> **EXAMPLE**
>
> You have withdrawn from a math course after mid-term due to low grades. Instead of waiting until next semester, attend a math lab or seek a tutor and learn Chapters 1, 2, and 3 *to perfection.* Also use this time to improve your math study skills. You will enter the same math course next semester with excellent math knowledge and study skills. In fact, you can make an A on the first test and complete the course with a high grade. Does this sound far-fetched? It may, but I know hundreds of students who have used this learning procedure and passed their math courses instead of failing them again and again.

Finding Your Best Instructor

Finding an instructor who best matches your learning style can be a difficult task. Your learning style is important; your learning style is how you best acquire information.

> **EXAMPLE**
>
> Auditory learners do better when hearing the information over and over again instead of carefully reading the information. If an auditory learner is taught by a visual-style instructor who prefers that students read materials on their own and who prefers working problems instead of describing them, the mismatch could cause the student to do worse than if the student were taught by an auditory instructor.

Most students are placed in their first math course by an academic advisor. Usually, academic advisors know who are the most popular and least popular math instructors. However, advisors can be reluctant to discuss teacher popularity. And, unfortunately, students may want the counselor to devise a course schedule based on the student's time limits instead of teacher selection.

To learn who are the best math instructors, ask the academic advisor which math instructor's classes fill up first. This does not place the academic advisor in the position of making a value judgment; neither does it guarantee the best instructor. But it will increase the odds in your favor.

Another way to acquire a good math instructor is to ask your friends about their current and previous math instructors. However, if another student says that an instructor is excellent, make sure your learning style matches your friend's learning style.

Ask your friend, Exactly what makes that instructor so good? Then compare the answer with how you learn best. If you have a different learning style than your friend, look for another instructor, or ask another friend whose learning style more closely matches your own.

To obtain the most from an instructor, discover your best learning style and match it to the instructor's teaching style. Most learning centers or student personnel offices will have counselors who can measure and explain your learning style.

Interview or observe the instructor while the instructor is teaching. This process is time consuming, but it is well worth the effort!

Once you have found your best instructor, do not change. Remain with the same instructor for every math class whenever possible.

Whether your learning styles are a natural fit for learning math or not, you still own the responsibility to be smart in selecting your study system for math class. You can be successful!

REMEMBER
The first step in becoming a better math student is knowing your learning strengths and challenges. Now you can focus on what you need to improve.

SECTION 2.5   REVIEW

1.   Give four examples of how you can review math course materials.

Example 1: _____

_____

_____

Example 2: _____

_____

_____

Example 3: _____

_____

_____

Example 4: _____

_____

_____

2.   What is the advantage of taking math courses back to back?

_____

_____

_____

3. Give two examples of how to find your best math instructor.

Example 1: _____

_____

_____

_____

Example 2: _____

_____

_____

_____

CHAPTER 2 REVIEW

1. _____ _____ _____ refers to how much math you knew before entering your current math course. (Answers on page 14)

2. _____ _____ _____ are the learning skills you possess that affect your grades. (Answers on page 14)

3. Students who talk their instructors into moving them up a course level usually _____ that course. (Answer on page 15)

4. The _____ of tutors is a major part of the math lab or learning resource center.

5. Affective student characteristics account for about _____ percent of your grade.

6. _____ learners look for facts and ask experts for their advice.

7. Hands-on learners are called _____ learners.

8. Assessing your math learning _____ and _____ will help you improve your studying and learning.

9. Students having previous difficulty in mathematics should start tutorial sessions the _____ week class begins.

10. Once you find your best math instructor, _____ _____ change instructors for your next math course.

11. What is the most important information you learned from this chapter?

    _____

12. How can you immediately use it?

    _____

DAN'S TAKE REVIEW

1. What was Dan's first mistake in putting off taking a math course until his senior year?

    _____

    _____

2. What were the potential consequences of his decision-making process?

    _____

    _____

3. How could he have avoided the issue, and how can you avoid this issue in the future?

    _____

    _____

How to Reduce Math Test Anxiety

CHAPTER 3

In this chapter you will first learn about math anxiety and then explore test anxiety. They are different issues that can affect each other. It is best to learn about them as separate entities first.

Math anxiety is a common problem for many high school, college, and university students. It is especially difficult for students in developmental courses, who normally have more math anxiety than other students. However, there are students in higher-level math courses who also struggle with this problem. It is very common for students to have anxiety only about math and not in their other subjects.

Most students think that math anxiety affects them only when they are taking tests, but it also affects other areas. It can affect the way you do your homework, the way you learn in the classroom or through distance learning courses, and the way you choose a career. Students who have math anxiety may procrastinate in doing their homework or put off sitting down and completing an online lesson. This can lead to math failure. Students also select majors based on the amount of math that is required, which could lead to lower-paying or dissatisfying careers. However, most students with math anxiety meet it face to face during tests, experiencing test anxiety as well.

Mild test anxiety can be a motivational factor in that it can make students properly prepare for a test. However, high test anxiety can cause major problems in both learning and test taking, as students avoid studying for the test when anxiety begins to afflict their thought processes. Reducing test anxiety is the key for many students to become successful in math. Such students need to learn the causes of test anxiety and how to reduce the test anxiety that affects their learning and grades.

Several techniques have proven helpful in reducing both math anxiety and math test anxiety. However, reducing them does not guarantee good math grades. It must be coupled with effective study skills and a desire to do well in math.

DAN'S TAKE

In 10th grade, I achieved the lowest test score of my life. The class: trigonometry. The grade: well, you could count my score on one hand.

As if this weren't embarrassing enough, when my teacher handed the test back to me, the entire group sitting at my table took notice before I could manage to stuff the exam underneath my textbook. But then, how could they miss it? The paper was so thoroughly covered with thick, radiating red-pen marks that it may well have been visible to the astronauts on the International Space Station.

But before you judge me, let's follow my thought process before the exam:

1. I am bad at math. I do not get it.
2. This test is the third in the semester and I have already failed the previous two.
3. If I failed the other tests, what are the chances that I'll pass the third?
4. Conclusion: there is absolutely no point in studying. I can't possibly do any worse right?

I was wrong. Horribly wrong. I *could* do worse . . . and I did. For one great night of videogames and television reruns, I had backed myself into a corner that nearly sunk my academic future. What's more, I nearly lost my spot on the basketball team as my midterm GPA was within a 10th of a point of making me ineligible.

When I chose not to study for this particular exam, I was attempting to avoid the shame of trying *and still* failing. I somehow justified the act of failing without studying as a smaller blow to my pride than knowing that I had put in the proper amount of time and effort and still come up flat.

What I did not realize at the time was that I was simply looking for an excuse to cover up the real issue behind my poor math grades: math anxiety. I convinced myself that I was just bad at math. Even a simple thought about that subject was enough to send me into a spiral of self-doubt and discomfort. What I did not know was that by merely thinking these thoughts, I was setting myself up to fail from the very beginning. My fear made me "draw blanks" during my exams and I often wound up failing.

It was not until I went to a tutor that I realized I could get over these fears. I had to stop using the excuse "I am just bad at math." With a little help, I pulled my final grade up to a C. It turned out I was not bad at math at all—I was merely afraid of it!

3.1   Understanding Math Anxiety

Definitions of Math Anxiety

Math anxiety is a relatively new concept in education. During the 1970s, certain educators began referring to *mathophobia* or *mathemaphobia* as a possible cause for children's unwillingness to learn math. Math anxiety is an extreme emotional and/or physical reaction to a very negative attitude toward math. There is a strong relationship between low math confidence and high math test anxiety (Fox, 1977).

Math anxiety is the feeling of tension and anxiety that interferes with the manipulation of numbers and the solving of math problems during tests (Richardson and Suinn, 1973). Math anxiety is a state of panic, helplessness, paralysis, and mental disorganization that occurs in some students when they are required to solve math problems. This discomfort varies in intensity and is the outcome of numerous experiences students have had in their past learning situations (Tobias, 1978).

Today, math anxiety is accepted as one of the major problems students have in completing their math courses. It is real, but it can be overcome.

Types of Math Anxiety

Math anxiety can be divided into three separate anxieties: math test anxiety, numerical anxiety, and abstraction anxiety. Math test anxiety involves anticipation of, completion of, and feedback on math tests. Numerical anxiety refers to everyday situations requiring people to work with numbers and perform arithmetic calculations. Numerical anxiety can also include students who are trying to figure out the amount of a tip, thinking about mathematics, doing math homework, or listening to or seeing math instruction. Abstraction anxiety involves working with variables and mathematical concepts used to solve equations. Students can have all three math anxieties or only one anxiety. Most often, the students I have worked with have had math test anxiety and abstraction anxiety. These students don't have any anxiety working with numbers, but once they start learning algebra, they develop both conditions. This may have happened in high school or college.

REMEMBER
When people try to overcome anxiety of any kind, including math anxiety, they must seek advice and support. It is much easier with support!

The Causes of Math Anxiety

Since math anxiety is a learned condition, its causes are unique to each student, but they are all rooted in individuals' past experiences. Bad experiences in elementary school are one of the most common sources of students' math anxiety: coming in last in math races at the blackboard, watching classmates finish problems twice as fast as they do, teachers saying, "That's okay, you just aren't good in math, you are better in English," and classmates and teachers calling them stupid. These words and experiences remain with people; they can still hear the words and eventually begin telling themselves the same thing. When these students walk into a classroom or open a math book, or take a test, these "mental tapes" play in their minds. When asked, many students indicate that they were made fun of when trying to solve math problems at the chalkboard. When they could not solve the problems, the teacher and/or students would call them "stupid."

Teacher and peer embarrassment and humiliation become the conditioning experiences that cause some students' math anxiety. Over the years, this math anxiety is reinforced and even increases in magnitude. In fact, many math anxious students—now 30, 40, and 50 years old—still have extreme fear about working math problems on the board. One 56-year-old indicated that he had a great deal of fear that the instructor would call him to the board. Even if he knew how to do the problem, displaying that knowledge to his peers was very difficult. Some students have said that they absolutely refused to go to the board.

Being embarrassed by family members can also cause math anxiety. According to students who have been interviewed on the matter, their parents tried to help them with math and this sometimes led to serious trauma. These students claim that the tutoring from their guardians, mainly their fathers, often resulted in scolding when they were not able to complete the problems. One student reported that his father hit him every time he got a multiplication problem wrong. Brothers and sisters can also tease one another about being "dumb" in math. This is particularly true of boys telling girls that they cannot do math. When people hear these statements often enough, they may start to believe them and associate these bad feelings with the word *math*. So, for students many years later, just hearing the word "math" triggers a response of anxiety, consciously or unconsciously recalling the bad feelings, and becoming uneasy.

A good example of this is a student I worked with who had completed her B.S. degree 15 years ago at a college that did not require much math. She was returning to college to be an elementary school teacher, which required her to take math and a math placement test. As soon as I mentioned that she had to take math, she said, "I can't do math and I will have to wait a few days to get psychologically ready to take the math placement test." She indicated that her old feeling of not being able to do math rushed through her and she almost had an anxiety attack. This is an extreme case, but a true example, of math anxiety. In most cases, math anxiety is not this bad, but it is disruptive enough to cause learning and testing problems.

In order to overcome math anxiety, it is necessary to identify when it first started and what caused it. Math anxiety usually starts from one or a series of events in a student's past that were negative. Negative feelings were imbedded in memories and then anxiety started whenever the student was placed in similar situations.

Answer the questions on the Exploring Past Math Experiences Worksheet.

If some of the memories are tough ones, just make a short note of them right now. Don't dwell on them, because they are in the past. This can be frustrating for some students, but it is worth working through it. You will realize that math anxiety is usually a result of events, and does not occur because you are "dumb" or have other personal flaws.

If you have had a very positive experience in the past when you were studying math, brainstorm answers to the appropriate questions. You may want to keep in mind friends you know who struggle with math and think about the other questions.

This brainstorming will help you write your math autobiography assignment at the end of this chapter.

SOME FACTS TO REMEMBER

- Math anxiety is usually a result of past experiences that were negative. As a result of these experiences, any time a person is put in a similar situation, he or she gets anxious.
- Math anxiety is not related to how smart a person is.
- Math anxiety is a learned condition; therefore, in most cases, it can be unlearned or at least managed.
- A person must be willing to change and find strategies to practice continually that will help manage math anxiety.

EXPLORING PAST MATH EXPERIENCES WORKSHEET

At the end of this chapter you will write a "Math Autobiography" to help you explore why you might have math anxiety. This knowledge is the first step in learning how to manage or overcome the anxiety. This worksheet will help you prepare for writing the autobiography. Answer the following questions. For some students, these questions will be easy to answer; for others, they will be more difficult because their past experiences are quite negative. The more honest you can be, the more you will be able to confront and work on any math anxiety.

1. When was the first time you were nervous or anxious about learning math? (Do not include taking tests.) Was it in elementary, middle, or high school? Can you recall a specific incident? If so, briefly describe it. It could have taken place in school or at home.

   Examples: Getting scolded by a teacher or parent for a poor homework grade. Losing a math race on the board. Going to recess late because you did not get your math work done in time. Older sibling always comparing math grades with you.

   _____

   _____

2. When was the last time you told yourself that you couldn't do math or were just not a math type?

   _____

   _____

3. Even though you might have math anxiety, there were probably some positive experiences associated with math in the past. Try to think of at least one of them.

   Examples: A good grade. Completing an assignment in class. A parent or teacher praising you for your math work.

   _____

   _____

4. If you have math anxiety, take heart. Most students can learn to minimize it and control it. Write a statement declaring that you will set a goal to learn to manage your math anxiety this semester.

   _____

   _____

- -

How Math Anxiety Affects Learning

Math anxiety can cause learning problems in several ways. It can affect how you do your homework and your participation in the classroom and in study groups. Let's first start by looking at how math anxiety could affect your homework. Students with high math anxiety may have difficulty starting or completing their homework.

Doing homework reminds some students of their learning problems in math. More specifically, it reminds them of their previous math failures, which causes further anxiety. This anxiety can lead to total avoidance of homework or "approach-avoidance" behavior.

Total homework avoidance is called *procrastination*. The very thought of doing their homework causes these students anxiety, which causes them to put off tackling their homework. This makes them feel better for a short amount of time—*until test day.*

Math anxiety can also affect your classroom participation and learning. Usually students with math anxiety are afraid to speak out in class and ask questions. They remember that in the past they were made fun of when they gave incorrect answers. They are also afraid of asking a question that others, including the teacher, will consider "dumb." So they sit in class fearful of being asked a question, looking like they understand the lecture so they will not be called on. They also take a lot of notes even though they don't understand them, in order to give the illusion of comprehension. If you are one of these students, these are hard habits to break. However, these habits may cause you to be unsuccessful in your math class. Here are some suggestions that may help you break these habits:

1. Make an appointment to talk to your math instructor. Math instructors want you to talk to them. When I do my consulting around the country, one of the major complaints I get from math instructors is that the students don't come and see them. Make an appointment to see your math instructor before the first major test to discuss your math history and to ask for learning suggestions. In most cases it is easier to talk to the instructor before you get your first grade.

2. Before class, ask the instructor to work one homework problem. You might want to write the problem on the board before the instructor arrives. This is less stressful because you are not asking the question in front of the whole class. In fact, one of my good friends, Dr. Mike Hamm, suggests that his students put the problems they do not know how to solve on the board before class. Other students go to the board and solve the problems. Dr. Hamm then solves the ones the students cannot solve.

3. Prepare one question from your homework and ask it within the first 15 minutes of class. Instructors are more likely to answer questions in the first part of class when they have

time instead of the end of class when time is running out.

4. Ask a question to which you already know the answer. The question will begin discussion with the instructor and other students when you can contribute since you understand the problem. It will set you up for a positive experience.

5. Use email to send questions to your instructor. This way, you can still get the answer with very little anxiety.

By using these suggestions, you can reduce your math anxiety and learn more mathematics. A question unanswered could be a missed test question.

REMEMBER

The instructor's job is to answer your questions, and you are paying for the course.

Math-anxious students sometimes avoid doing additional math outside of the classroom. They avoid study groups and supplemental instruction. It is like asking a person with hydrophobia (fear of water) to take a vacation at the beach. However, a person with hydrophobia can go to the beach and enjoy himself or herself and not get wet. In other words, students can still attend study groups and supplemental instruction and just listen. When they are ready to get their feet wet, they can ask a few questions. Don't let these great opportunities go by.

Math anxiety can affect how you learn mathematics. It can be overcome with your effort. You don't have to live in the past with your math fears. Today is a new day, and you can change how math affects you. The next step is to understand how test anxiety can affect your demonstration of math knowledge.

SECTION 3.1   REVIEW

1. List and explain the three types of math anxiety.

   First Type: _____

   Second Type: _____

   Third Type: _____

2. List and explain two causes of math anxiety.

   First Cause: _____

   Second Cause: _____

3. List three ways math anxiety can cause learning problems.

   First Way: _____

   Second Way: _____

   Third Way: _____

4. List and explain three ways students can manage or avoid the effects of math anxiety.

   First Way: _____

   Second Way: _____

   Third Way: _____

3.2 How to Recognize Test Anxiety

Test anxiety has existed for as long as tests have been used to evaluate student performance. Because it is so common and because it has survived the test of time, test anxiety has been carefully studied over the last 50 years. Pioneering studies indicate that test anxiety generally leads to low test scores. At the University of South Florida (Tampa), Dr. Charles Spielberger investigated the relationship between test anxiety and intellectual ability. The study results suggested that anxiety coupled with high ability can improve academic performance; but anxiety coupled with low or average ability can interfere with academic performance. That is:

anxiety + high ability = improvement

anxiety + low or average ability = no improvement

> **EXAMPLE**
>
> Students with average ability and low test anxiety had better performance and higher grades than did students with average ability and high test anxiety. However, there are students who make good grades, take calculus, and still have test anxiety.

Test anxiety is a *learned* response; a person is not born with it. An environmental situation brings about test anxiety. The good news is that because it is a learned response, it can be *unlearned.* Test anxiety is a special kind of general stress. General stress is considered to be "strained exertion," which can lead to physical and psychological problems.

Math Test Anxiety

There are several definitions of test anxiety. One definition states, "Test anxiety is a conditioned emotional habit to either a single terrifying experience, recurring experience of high anxiety, or a continuous condition of anxiety" (Wolpe, 1958).

Another definition of test anxiety relates to the educational system. The educational system develops evaluations that measure one's mental performance, and this creates test anxiety. This definition suggests that test anxiety is the *anticipation* of some realistic or nonrealistic situational threat (Cattell, 1966). The "test" can be a research paper, an oral report, work at the chalkboard, a multiple-choice exam, a written essay, or a math test.

Math test anxiety is a relatively new concept in education. *Ms.* magazine published "Math Anxiety: Why Is a Smart Girl Like You Counting on Your Fingers?" (Tobias, 1976) and coined the phrase *math anxiety.* During the 1970s, other educators began referring to *mathophobia* or *mathemaphobia* as a possible cause of children's unwillingness to learn math. Additional studies on the graduate level discovered that math anxiety was common among adults as well as children.

One of my students once described math test anxiety as "being in a burning house with no way out." No matter how you define it, math test anxiety is real, and it affects millions of students.

Why Math Tests Create Anxiety

It has been shown that math anxiety exists among many students who usually do not suffer from other tensions. Counselors at a major university reported that one-third of the students who enrolled in behavior therapy programs offered through counseling centers had problems with math anxiety (Suinn, 1988).

Educators know that math anxiety is common among college students and is more prevalent in women than in men. They also know that math anxiety frequently occurs in students with poor high school math backgrounds. These students were found to have the greatest amount of anxiety.

Approximately half of the students in college prep math courses (designed for students with inadequate high school math backgrounds or low placement scores) could be considered to have math anxiety. However, math anxiety also occurs in students in high-level math courses, such as college algebra and calculus.

Educators investigating the relationship between anxiety and math have indicated that anxiety contributes to poor grades in math. They have also found that simply *reducing* math test anxiety does not guarantee higher math grades. Students often have other problems that affect their math grades, such as poor study skills, poor test-taking skills, or poor class attendance.

SECTION 3.2 REVIEW

1. What is your personal definition of test anxiety?

_____

2. What type of student does math anxiety affect most?

_____

3.3 The Causes of Test Anxiety

The causes of test anxiety can be different for each student, but they can be explained by seven basic concepts. See Figure 3.

The causes of math test anxiety can be different for each student. It could possibly have first occurred in middle school or high school. However, for many students it first occurs in college when passing tests is the only way to pass the course. Homework and extra credit in most college courses don't count toward your grade. Now students must have a passing average and in some cases must pass the departmental final exam. Additional pressure also exists because not passing algebra means you won't graduate and you might not get the job you want. As you can see, there are more reasons to have math test anxiety in college than in high school.

If you have math test anxiety, I want you to try to remember the first time it surfaced. Was it in middle school, high school, or college? Can you recall a specific incident? Was it your first algebra test? Was it your first math test after being out of school for a long time? Was it after you decided to get serious about college? Was it a college algebra course that was required for your major? Was it an instructor who told you that if you didn't pass the next math test you would fail the course? Was it when you needed to pass the next math test to pass the course so you could keep your financial aid? Was it your children asking you why you failed your last math test? If you cannot remember a specific incident when you had test anxiety, do you expect to have any major anxiety on the next test you take?

FIGURE 3 Causes of Test Anxiety

- Test anxiety can be a learned behavior resulting from the expectations of parents, teachers, or other significant people in the student's life.
- Test anxiety can be caused by the association between grades and a student's personal worth.
- Test anxiety develops from fear of alienating parents, family, or friends due to poor grades.
- Test anxiety can stem from a feeling of lack of control and an inability to change one's life situation.
- Test anxiety can be caused by a student's being embarrassed by the teacher or other students when trying to do math problems.
- Test anxiety can be caused by timed tests and the fear of not finishing the test, even if one can solve all the problems.
- Test anxiety can be caused by being put in math courses above one's level of competence.
- What is (are) the cause(s) of your *test* anxiety?

Students with math test anxiety also have had positive experiences with taking math tests. Now try to remember your first positive experience with taking a math test. Was it in middle school, high school, or college? Was it after studying many hours for the test? Now think back to your last positive experience with a math test. How did it make you feel?

Since we have already explored your experiences with taking math tests, let's look at some of the direct causes of your math test anxiety. If you do have test anxiety, what is the main cause? If you don't know, then review the seven causes of test anxiety in Figure 3. Does one of these reasons fit you? If you don't have test anxiety, what would be a reason that could cause test anxiety?

Like you did before, brainstorm answers to the questions listed in the previous three paragraphs. You will also use this brainstorming to help you write your autobiography.

REMEMBER

For some students, just writing about their previous math history helps them.

SECTION 3.3 REVIEW

1. List and describe five causes of test anxiety.

First Cause: _____

_____

_____

Second Cause: _____

_____

_____

Third Cause: _____

_____

_____

Fourth Cause: _____

_____

_____

Fifth Cause: _____

_____

_____

2. Put into your own words your cause(s) of test anxiety. If you don't have test anxiety, list what you think is the major cause of test anxiety.

_____

_____

3.4 The Different Types of Test Anxiety

The two basic types of test anxiety are emotional anxiety (educators call this *somatic* anxiety) and worry anxiety (educators call this *cognitive* anxiety). Students with high test anxiety have *both* emotional and worry anxiety.

Signs of emotional anxiety are upset stomach; nausea; sweaty palms; pain in the neck; stiff shoulders; high blood pressure; rapid, shallow breathing; rapid heartbeat; and general feelings of nervousness. As anxiety increases, these feelings intensify. Some students even run to the bathroom to throw up or have diarrhea.

Even though these *feelings* are caused by anxiety, the physical response is real. These feelings and physical inconveniences can affect your concentration and your testing speed, and they can cause you to "draw a blank."

Worry anxiety causes the student to think about failing the test. These negative thoughts can happen either before or during the test. This negative "self-talk" causes students to focus on their anxiety instead of recalling math concepts.

The effects of test anxiety range from a "mental block" on a test to avoiding homework. One of the most common side effects of test anxiety is getting the test and immediately forgetting information that you know. Some students describe this event as having a mental block, going blank, or finding that the test looks like a foreign language.

After 5 or 10 minutes into the test, some of these students can refocus on the test and start working the problems. They have, however, lost valuable time. For other students, anxiety persists throughout the test and they cannot recall the needed math information. It is only after they walk out the door that they can remember how to work the problems.

Sometimes math anxiety does not cause students to go blank but slows down their mental processing speed. This means it takes longer to recall formulas and concepts and to work problems. The result is frustration and loss of time, leading to *more* anxiety. Since, in most cases, math tests are *speed* tests (those in which you have a certain amount of time to complete the test), you may not have

enough time to work all the problems or to check the answers if you have mentally slowed down. The result is a lower test score, because even though you know the material, you do not complete all of the questions before test time runs out.

Not using all of the time allotted for the test is another problem caused by test anxiety. Students know that they should use all the test time to check their answers. In fact, math is one of the few subjects in which you can check test problems to find out if your work is correct. However, most students do not use all of the test time, and this results in lower test scores. Why does this happen?

Students with high test anxiety do not want to stay in the classroom. This is especially true of students whose test anxiety increases as the test progresses. The test anxiety gets so bad that they would rather leave early and receive a lower grade than stay in that "burning house."

Students have another reason for leaving the test early: the fear of what the instructor and other students will think about them for being the last one to hand in the test. These students refuse to be the last ones to finish the test because they think that the instructor or other students will think they are "dumb." This is middle school thinking, but the feelings are still real—no matter what the age of the student. These students do not realize that some students who turn in their tests first fail, whereas many students who turn in their tests last make As and Bs.

Another effect of test anxiety relates to completing homework assignments. Students who have high test anxiety may have difficulty starting or completing their math homework. Doing homework reminds some students of their learning problems in math. More specifically, it reminds them of their previous math failures, which causes further anxiety. This anxiety can lead to total homework avoidance or "approach-avoidance" behavior.

Total homework avoidance is called *procrastination*. The very thought of doing their homework causes these students anxiety, which causes them to put off tackling their homework. This makes them feel better for a short amount of time—*until test day.*

EXAMPLE

Some students begin their homework and work some problems successfully. They then get stuck on a problem that causes them anxiety, so they take a break. During their break the anxiety disappears until they start doing their homework again. Continuing their homework causes more anxiety, which leads to another break. The breaks become more frequent. Finally, the student ends up taking one long break and not doing the homework. Quitting, to them, means *no more anxiety* until the next homework assignment.

FIGURE 4   The 12 Myths About Test Anxiety

1. Students are born with test anxiety.
2. Test anxiety is a mental illness.
3. Test anxiety cannot be reduced.
4. Any level of test anxiety is bad.
5. All students who are not prepared have test anxiety.
6. Students with test anxiety cannot learn math.
7. Students who are well prepared will not have test anxiety.
8. Very intelligent students and students taking high-level courses, such as calculus, do not have test anxiety.
9. Attending class and doing homework should reduce all test anxiety.
10. Being told to relax during a test will make a student relaxed.
11. Doing nothing about test anxiety will make it go away.
12. Reducing test anxiety will guarantee better grades.

The effects of math test anxiety can be different for each student. Students can have several of the mentioned characteristics that can interfere with math learning and test taking. However, there are certain myths about math that each student needs to know. Review Figure 4, The 12 Myths About Test Anxiety, to see which ones you believe. If you have test anxiety, which of the items listed is (are) true for you?

SECTION 3.4   REVIEW

1. List and describe the two basic types of test anxiety.

   First Type: _____

   Second Type: _____

2. List two reasons students leave the test room early instead of checking their answers.

   First Reason: _____

   Second Reason: _____

3. List 6 of the 12 myths about test anxiety that you most believed.

   First Myth: _____

   Second Myth: _____

   Third Myth: _____

   Fourth Myth: _____

   Fifth Myth: _____

   Sixth Myth: _____

3.5 How to Reduce Test Anxiety

To reduce math test anxiety, you need to understand both the relaxation response and how negative self-talk undermines your abilities.

Relaxation Techniques

The relaxation response is any technique or procedure that helps you to become relaxed. It will take the place of an anxiety response. Someone simply telling you to relax or even telling yourself to relax, however, does little to reduce your test anxiety. There are both short-term and long-term relaxation response techniques that help control emotional (somatic) math test anxiety. These techniques will also help reduce worry (cognitive) anxiety. Effective *short-term* techniques include the tensing and differential relaxation method, the palming method, and deep breathing.

Short-Term Relaxation Techniques

The Tensing and Differential Relaxation Method The tensing and differential relaxation method helps you relax by tensing and relaxing your muscles all at once. Follow these procedures while you are sitting at your desk before taking a test:

1. Put your feet flat on the floor.
2. With your hands, grab underneath the chair.
3. Push down with your feet and pull up on your chair at the same time for about 5 seconds.
4. Relax for 5 to 10 seconds.
5. Repeat the procedure two or three times.
6. Relax all your muscles except the ones that are actually used to take the test.

The Palming Method The palming method is a visualization procedure used to reduce test anxiety. While you are at your desk before or during a test, follow these procedures:

1. Close and cover your eyes using the center of the palms of your hands.
2. Prevent your hands from touching your eyes by resting the lower parts of your palms on your cheekbones and placing your fingers on your forehead. Your eyeballs must not be touched, rubbed, or handled in any way.
3. Think of some real or imaginary relaxing scene. Mentally visualize this scene. Picture the scene as if you were actually there, looking through your own eyes.
4. Visualize this relaxing scene for 1 to 2 minutes.
5. Open your eyes and wait about 2 minutes and visualize the scene again. This time also imagine any sounds or smells that can enhance your scene. For example, if you are imagining a beach scene, imagine the sound of the waves on the beach and the smell of the salty air. This technique can also be completed without having your hands over your eyes by just closing your eyes.

Practice visualizing this scene for several days before taking a test and the effectiveness of this relaxation procedure will improve.

Deep Breathing Deep breathing is another short-term relaxation technique that can help you relax. Proper breathing is a way to reduce stress and decrease test anxiety. When you are breathing properly, enough oxygen gets into your bloodstream to nourish your body and mind. A lack of oxygen in your blood contributes to an anxiety state that makes it more difficult to react to stress. Proper deep breathing can help you control your test anxiety.

Deep breathing can replace the rapid, shallow breathing that sometimes accompanies test anxiety, or it can prevent test anxiety. Here are the steps in the deep-breathing technique:

1. Sit straight up in your chair in a good posture position.
2. Slowly inhale through your nose.
3. As you inhale, first fill the lower section of your lungs and work your way up to the upper part of your lungs.
4. Hold your breath for a few seconds.

5. Exhale slowly through your mouth.

6. Wait a few seconds and repeat the cycle.

7. Keep doing this exercise for 4 or 5 minutes. This should involve going through about 10 breathing cycles. Remember to take two normal breaths between cycles. If you start to feel lightheaded during this exercise, stop for 30 to 45 seconds and then start again.

8. Throughout the entire exercise, make sure you keep breathing smoothly and in a regular rhythm without gulping air or suddenly exhaling.

9. As an extra way to improve your relaxation, say "relax" or "be calm" to yourself as you exhale. This can start a conditioned response that can trigger relaxation when you repeat the words during anxious situations. As you keep practicing, this conditioned response will strengthen. Practice is the key to success.

You need to practice this breathing exercise for several weeks before using the technique during tests. If you don't practice this technique, it will not work. After taking your first test, keep doing the exercise several times a week to strengthen the relaxation response.

The CD *How to Reduce Test Anxiety* further explains test anxiety and discusses these and other short-term relaxation response techniques. Short-term relaxation techniques can be learned quickly but are not as successful as the long-term relaxation technique. Short-term techniques are intended to be used while learning the long-term technique.

Long-Term Relaxation Technique

The cue-controlled relaxation response technique is the best long-term relaxation technique. It is presented on the CD *How To Reduce Test Anxiety*. Cue-controlled relaxation means you can induce your own relaxation based on repeating certain cue words to yourself. In essence, you are taught to relax and then silently repeat cue words, such as, "I am relaxed." After enough practice, you can relax during math tests. The cue-controlled relaxation technique has worked with thousands of students. For a better understanding of test anxiety and how to reduce it, listen to *How to Reduce Test Anxiety.*

Negative Self-Talk

According to cognitive psychologists, self-talk is what we say to ourselves as a response to an event or situation. This self-talk determines our feelings about that event or situation. Sometimes we say it so quickly and automatically that we don't even hear ourselves. We then think it is the situation that causes the feeling, but in reality it is our interactions or thoughts about the experience that are controlling our emotions. This sequence of events can be represented by the following time line:

External Events
(Math test)

Interpretation of Events and Self-Talk
(How you feel about the test
and what you are telling yourself)

Feelings and Emotions
(Happy, glad, angry, mad, upset)

In most cases this means that you are responsible for how and what you feel. You can engage in positive self-talk or negative self-talk about taking a math test. Yes, some students practice positive self-talk about math tests and see each test as a challenge and something to accomplish that makes them feel good, while others see it as an upsetting event that leads to anger and anxiety. In other words, you are what you tell yourself.

Negative self-talk is a form of worry (cognitive) anxiety. This type of worrying can interfere with your test preparation and can keep you from concentrating on the test. Worrying can motivate you to study, but too much worrying may prevent you from studying at all.

Negative self-talk is defined as the negative statements you tell yourself before and during tests. Negative self-talk causes students to lose confidence and to give up on tests. Furthermore, it can give you an inappropriate excuse for failing math and cause you to give up on learning math.

Examples of Negative Self-Talk

"It doesn't matter how hard I work. I am going to fail the class no matter what."

"I have already failed this class before. There is no way I can pass it this time, either."

"I do not get these problems. I am going to fail this test and flunk the course."

"I knew how to do these problems not 15 minutes ago, and now . . . nothing! What is up with me, I always do this when I take math tests."

"If I can't even pass this easy test, how can I expect to pass the math classes I need to graduate?"

Students who have test anxiety are prone to negative self-talk. Test anxiety can be generated or heightened by repeatedly making statements to yourself that usually begin with What if? For example, What if I fail the test? or What if I fail this class again? These what-if statements generate more anxiety, which can cause students to feel sick. These statements tell them to be anxious. Some other aspects of self-talk are as follows:

- Self-talk can be in telegraphic form with short words or images.
- Self-talk can be illogical, but at the time, the person believes it.
- Negative self-talk can lead to avoidance such as not taking a test or skipping classes.
- Negative self-talk can cause more anxiety.
- Negative self-talk can lead to depression and a feeling of helplessness.
- Negative self-talk is a bad habit that can be changed.

There are different types of negative self-talk. If you have negative self-talk, then review the following types and see which one fits you best. You may use a combination of them.

1. *The Critic* is the person inside us who is always trying to put us down. It constantly judges our behaviors and finds fault even if there is none. It jumps on any mistake and exaggerates it to cause more anxiety. The Critic puts us down for mistakes on tests and blames us for not controlling our anxiety. The Critic reminds us of previous comments from real people who have criticized us. It compares us with other students who are doing better in the class. It loves to say, "That was a stupid mistake!" or "You are a total disappointment. You can't pass this math class like everyone else can!" The Critic's goal is to promote low self-esteem.

2. *The Worrier* is the person inside us who looks at the worst-case scenario. It wants to scare us with ideas of disaster and complete failure. When it sees the first sign of anxiety, it "blows it out of proportion" to the extent that we will not remember anything and totally fail the test. The Worrier creates more anxiety than normal. The Worrier anticipates the worst, underestimates our ability, and sees us not only failing the test but also "failing life." The Worrier loves to ask What if? For example, What if I fail the math test and don't graduate? or What if I can't control my anxiety and throw up in math class? The goal of the Worrier is to cause more anxiety so we will quit.

3. *The Victim* is the person inside us who wants us to feel helpless or hopeless. It wants us to believe that no matter what we do, we will not be successful in math. The Victim does not blame other events (poor schooling) or people (bad teachers) for our math failures. It blames us. It dooms us and puts us into a learned helpless mode, meaning that if we try to learn math, we will fail, or if we don't try to learn math, we will fail. So why try? The Victim likes to say, "I can't learn math." The goal of the Victim is to cause depression and make us stop trying.

4. *The Perfectionist* is similar to the Critic but is the opposite of the Victim. It wants us to do the best we can and will guide us into doing better. It tells us that we are not studying enough for math tests and that a B is not good enough and that we must make an A. In fact, sometimes an A is not good enough unless it is a score of 100%. So, if we make a B on a test, the Perfectionist says, "An A or a B is just like making an F." The Perfectionist is the hard-driving part of us that wants the best but cannot stand mistakes or poor grades. It can drive us to mental and physical exhaustion to make that perfect grade. It is not interested in

self-worth, just perfect grades. Students driven by the Perfectionist often drop a math course because they only have a B average. The Perfectionist loves to repeat, "I should have . . ." or "I must. . . ." The goal of the Perfectionist is to cause chronic anxiety that leads to burnout.

Review these types of personalities to see which one may fit you best. We may have a little of each one in us, but what is the dominant one for you in math? Now we can look at how to stop these negative thoughts.

Managing Negative Self-Talk

Students need to change their negative self-talk to positive self-talk without making unrealistic statements.

Positive self-statements can improve your studying and test preparation. During tests, positive self-talk can build confidence and decrease your test anxiety. These positive statements (see examples) can help reduce your test anxiety and improve your grades. Some more examples of positive self-statements are on the CD *How to Reduce Test Anxiety*. Before each test, make up some positive statements to tell yourself.

There are several ways to counter and control negative self-talk. Students can replace negative self-talk with positive statements and questions that make them think in a realistic way about the situation. Another way is to develop thought-stopping techniques to reduce or eliminate the negative thoughts. Try each way or a combination to see what works best for you.

Countering self-talk involves writing down and practicing positive statements that can replace negative statements. Students can develop their own positive statements. Some rules for writing positive statements are

1. Use the first-person present tense. For example, "I can control my anxiety and relax myself."

2. Don't use negatives in the statement. For example, don't say, "I am not going to have anxiety." Instead, say, "I will be calm during the test."

3. Have a positive and realistic belief in the statement. For example, say, "I am going to pass this test," instead of saying, "I am going to make

an A on this test" when you have not made an A on any of the tests.

The statements used to counter negative thoughts can be based on the type of negative self-talk we tend to engage in. The Critic, who puts us down by saying, "Your test anxiety is going to cause you to fail the test," can be countered with, "I have test anxiety, but I am learning to control it." The Worrier, who asks, "What if I fail the test?" can be countered with "So what? I will do better on the next test." The Victim, who thinks things are hopeless and says, "I will never be able to pass math," can be countered with, "I have studied differently for this test, and I can pass the math course." The Perfectionist, who says, "I need to make an A on the test or I will drop out of school," can be countered with, " I don't need to make an A to please anyone. All I need is to pass the math course to get the career I want." These are some of the examples of positive statements that can control anxiety.

Examples of Positive Self-Talk

"I might have failed the course last semester, but things are going to be different this time now that I have learned how to study math."

"Now that I know how to fight off test anxiety, there is no way I will panic during an exam like I did before."

"Just because I haven't had perfect results in the past doesn't mean that I am bad at math. This time, I know how to study math, so this course should not be a problem."

"If I put in the work, then I will pass with flying colors."

"I have done everything I needed to do to prepare for this test. If I can keep my composure during the test, and not get frustrated if I don't understand one or two problems, I should do all right."

"I will not let this one problem bother me. I have answered all the other problems correctly, so I cannot let this one make me miss the rest of the questions on the test."

"I will not hurry in order to be the first person to finish this test. I can go ahead and take my time to make sure I don't make any careless mistakes."

Thought-Stopping Technique

Some students have difficulty stopping their negative self-talk. These students cannot just tell themselves to eliminate those thoughts. These students need to use a thought-stopping technique to overcome their worry and become relaxed.

Thought stopping involves focusing on the unwanted thoughts and, after a few seconds, suddenly stopping those thoughts by emptying your mind. Using the command Stop! or a loud noise like clapping your hands can effectively interrupt the negative self-talk. In a homework situation, you may be able to use a loud noise to stop your thoughts, but don't use it during a test.

To stop your thoughts in the classroom or during a test, silently shout to yourself, "Stop!" or "Stop thinking about that!" After your silent shout, either relax yourself or repeat one of your positive self-talk statements. You may have to shout to yourself several times during a test or while doing homework to control negative self-talk. After every shout, use a different relaxation technique/scene or positive self-talk statement.

Thought stopping works because it interrupts the worry response before it can cause high negative emotions. During that interruption, you can replace the negative self-talk with positive statements or relaxation. However, students with high worry anxiety should practice this technique for three days to one week before taking a test. Contact your counselor if you have additional questions about the thought-stopping technique.

Writing Your Math Autobiography

An autobiography relates how you remember and feel about your past experiences. In addition, many autobiographies explore how these past feelings and experiences shape current life. While some people write autobiographies so that others can learn about their lives, many people write private autobiographies in order to understand what is going on in their lives.

Your final product will be a typed (or handwritten if your instructor approves) paper that summarizes your experiences learning math, how you feel about these experiences, and how they shape your current perspective on learning math.

SECTION 3.5   REVIEW

1. Describe your best short-term relaxation technique.

   _____

2. You can use the palming method by closing your eyes and visualizing a scene without putting your hands to your face. Describe a very relaxing scene that you could visualize. Make sure to include some sounds and visual images in your scene.

   _____

   _____

   _____

   _____

3. Practice your relaxation scene for 3 to 5 minutes for the next 5 days. List the times and dates you practiced your scene.

   Date: _____   Time: _____

   Date: _____   Time: _____

   Date: _____   Time: _____

Date: _____ Time: _____

Date: _____ Time: _____

4.   How does negative self-talk cause you to have text anxiety?

_____

_____

_____

_____

5.   Make up three positive self-talk statements that are not listed in this text.

Statement 1: _____

Statement 2: _____

Statement 3: _____

6.   Describe the thought-stopping technique in your own words.

_____

7.   What word or words will you use as your silent shout?

_____

8.   What will you do after your silent shout?

_____

9.   How does the thought-stopping technique work?

_____

CHAPTER 3   REVIEW

1. Reducing math test anxiety _____ _____ guarantee good math grades. (Answer on page 38)

2. Test anxiety is a _____ response; a person is not _____ with it. (Answers on page 37)

3. Math test anxiety involves _____, _____, and _____ of math tests.

4. One cause of test anxiety is that a student goes to the board to work a problem but is called _____ when he or she cannot work the problem.

5. The two basic types of test anxiety are _____ and _____.

6. The effects of test anxiety range from a "_____ _____" on a test to _____ homework.

7. The _____ and _____ _____ method helps you relax by tensing and relaxing your muscles all at once.

8. The palming method is a _____ procedure used to reduce text anxiety.

9. Negative self-talk is a form of _____ anxiety.

10. Thought stopping works because it interrupts the _____ response before it can cause _____ emotions.

What is the most important information you learned from this chapter?

_____

_____

_____

How can you immediately use it?

_____

_____

_____

DAN'S TAKE REVIEW

1. What was Dan's mistake in preparing for his math test?

_____

_____

_____

2. What were the potential consequences of his decision-making process?

   _____

   _____

   _____

3. How could he have avoided the issue, and how can you avoid this issue in the future?

   _____

   _____

   _____

How to Improve Your Listening and Note-Taking Skills

CHAPTER 4

Listening and note-taking skills in a math class are very important, since most students either do not read the math text or have difficulty understanding it. In most of your other classes, if you do not understand the lecture, you can read the text and get almost all the information. In math class, however, the instructor can usually explain the textbook better than the students can read and understand it.

Students who do not have good listening skills or note-taking skills will be at a disadvantage in learning math. Most math understanding takes place in the classroom. Students must learn how to take advantage of learning math in the classroom by becoming effective listeners, calculator users, and note takers.

51

DAN'S TAKE

My university became an "e-campus" my freshman year. Gone were the days of pencil and paper. Now they gave us laptops. The tapping of computer keys became a common soundtrack for class, as students tried frantically to type the last words of a PowerPoint slide into their notes before the professor hit the "next slide" button—or perhaps more accurately, as they instant messaged their friends across campus.

There is no question that students today have more distractions in the classroom than ever before. More campuses are using computers in the classroom, making paying attention nearly impossible. How can professors compete in a world with instant access to celebrity gossip rags and up-to-the-second BCS standings?

Yet math teachers can rejoice. It is very difficult to write math problems using a computer. Even when using online math courses, all the work has to be done on paper. This means there is no time for distraction.

On my particular campus, this made for a rather bizarre phenomenon. To many us, our computers were an extension of our bodies. They might well have been attached to our fingers. So math classes at my school presented an odd and old-school set of problems. The lack

of distraction actually became a distraction itself. If one of your hands is missing, well, that's the kind of thing a person takes notice of.

So to fill my diversionary void, I fell into a pattern of behavior that can only be described as idiotic. I cannot tell you how many times I would find myself doing something utterly pointless instead of taking notes. Things like flipping a pencil or staring at the cute sophomore sitting on the other side of the room.

Here is the problem with this scenario. Math is entirely based upon sequential learning. Two flips of a pencil and one love-stricken gaze might be enough to derail an entire lecture's worth of understanding. In my case, by the time I would refocus I'd be too far behind. This would require me to borrow someone else's notebook or figure out the correct processes through the homework itself.

The key to staying focused in class is to enter the room with a plan. The note-taking methods described in this chapter are invaluable. They really help you keep an organized notebook. It is important to be able to reference your notes easily during homework and remember which information is needed to answer each type of problem.

4.1   How to Become an Effective Listener

Becoming an effective listener is the foundation for good note taking. You can become an effective listener using a set of skills that you can learn and practice. To become an effective listener, you must prepare yourself both physically and mentally.

Sitting in the Golden Triangle

The physical preparation for becoming an effective listener involves *where you sit* in the classroom. Sit in the best area to obtain high grades, that is, in "the golden triangle of success." The golden triangle of success begins with seats in the front row facing the instructor's desk (see Figure 5).

Students seated in this area (especially in the front row) directly face the teacher and are most likely to pay attention to the lecture. This is a great seating location for visual learners. There is also

less tendency for them to be distracted by activities outside the classroom or by students making noise within the classroom.

The middle seat in the back row is another point in the golden triangle for students to sit, especially those who are auditory (hearing) learners. You can hear the instructor better because the instructor's voice is projected to that point. This means that there is less chance of misunderstanding the instructor, and you can hear well enough to ask appropriate questions.

By sitting in the golden triangle of success, you can force yourself to pay more attention during class and be less distracted by other students. This is very important for math students, because math instructors usually go over a point once and continue on to the next point. If you miss that point in the lesson, then you could be lost for the remainder of the class.

FIGURE 5   The Golden Triangle of Success

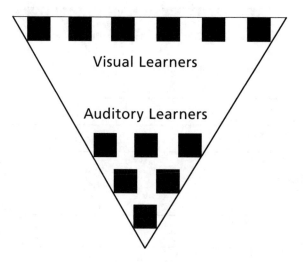

Back of the Classroom

This may seem obvious, but I know too many students who do the following if they have early morning classes. They roll out of bed, put on the closest clothes to them, brush their hair, maybe their teeth, and race to class. Don't do this. You won't wake up until the end of your class! Train yourself to be somewhere on campus to review your notes at least 15 minutes before class starts and follow some of the other strategies suggested in the next section. So, another physical way to be ready is to be awake and nourished!

Warming Up for Math Class

The mental preparation for note taking involves "warming up" before class begins and becoming an active listener. Just as an athlete must warm up before a game begins, you must warm up before taking notes. Warm up by

- reviewing the previous day's notes
- reviewing the reading material
- reviewing the homework
- preparing questions
- working one or two unassigned homework problems

This mental warm-up before the lecture allows you to refresh your memory and prepare pertinent questions, making it easier to learn the new lecture material.

How to Become an Active Listener

Becoming an active listener is the second part of the mental preparation for note taking.

Do not anticipate what the instructor is going to say or immediately judge the instructor's information before the point is made. This will distract you from learning the information.

> **EXAMPLE**
>
> Watch the speaker, listen for main ideas, and nod your head or say to yourself, "I understand" when agreeing with the instructor.

Expend energy looking for interesting topics in the lecture. When the instructor discusses information that you need to know, immediately repeat it to yourself to begin the learning process.

You can practice this exercise by viewing math DVD's and repeating important information. This is an especially good learning technique for auditory learners.

Ask appropriate questions when you have them. Sometimes if you are so confused you do not know what question to ask, still let the instructor know you are confused. Also, mark your notes where you got lost so that if the instructor doesn't have time to answer your question, you can visit during office hours to get an answer.

 REMEMBER
Class time is an intense study period that should not be wasted.

Listening and Learning

Some students think that listening to the instructor and taking notes are a waste of valuable time. Students too often sit in class and use only a fraction of their learning ability. Class time should be considered a valuable study period where you can listen, take notes, and learn at the same time. One way to do this is by memorizing important facts when the instructor is talking about material you already know. Another technique is to repeat to yourself the important concepts right after the instructor discusses them in class. Using class time to learn math is an efficient learning system.

SECTION 4.1 REVIEW

1. Explain how sitting in the golden triangle of success makes you a better listener.

_____

_____

_____

2. List five ways you can warm up before math class begins.

First Way: _____

Second Way: _____

Third Way: _____

Fourth Way: _____

Fifth Way: _____

3. Select your best warm-up process and try it before your next math class. What was your warm-up process, and how did it help you?

_____

_____

_____

_____

4. How can you listen and learn at the same time?

_____

_____

_____

_____

4.2 How to Become a Good Note Taker

Becoming a good note taker requires two basic strategies. One strategy is to be specific in detail. In other words, *copy* the problems down, step by step. The second strategy is to *understand* the general principles, general concepts, and general ideas.

Copying from the Board

While taking math notes, you need to copy each and every step of each problem even though you may already know every step. While in the classroom, you might understand each step, but a week later you might not remember how to do the problem unless you have written down all the steps. In addition, as you write down each step, you are memorizing it. Make sure to copy every step of each problem written on the board.

There will be times when you will get lost while listening to a lecture. Nevertheless, you should keep taking notes, even though you do not understand the problem. This will provide you with a reference point for further study. Put a question mark by each step that you do not understand; then, after class, review the steps you did not understand with the instructor, your tutor, or another student.

If you seem to think that many other students are lost, too, ask questions or let the instructor know.

Taking Notes

The goal of note taking is to take the least amount of notes and get the greatest amount of information on your paper. This could be the opposite of what most instructors have told you. Some instructors tell you to take down everything. This is not necessarily a good note-taking system, since it is very difficult to take precise, specific notes while trying to understand the instructor.

What you need to develop is a note-taking system in which you write the least amount possible and get the most information down while still understanding what the instructor is saying.

Developing an Abbreviation List

To reduce the amount of written notes, an abbreviation system is needed. An abbreviation system is your own system of reducing long words to shorter versions that you still can understand. By writing less, you can listen more and gain a better understanding of the material.

Figure 6 provides a list of abbreviations. Add your own abbreviations to this list. By using abbreviations as much as possible, you can obtain the same meaning from your notes and have more time to listen to the instructor.

> **EXAMPLE**
>
> When the instructor starts explaining the commutative property, you need to write "commutative property" out the first time. After that, use "com." You should develop abbreviations for all the most commonly used words in math.

FIGURE 6   Abbreviations

e.g.	(for example)
cf	(compare, remember in context)
N.B.	(note well, this is important)
\	(therefore)
∴	(because)
É	(implies, it follows from this)
>	(greater than)
<	(less than)
=	(equals, is the same)
≠	(does not equal, is not the same)
( )	(parentheses in the margin, around a sentence or group of sentences, indicates an important idea)
?	(used to indicate that you do not understand the material)
O	(a circle around a word may indicate that you are not familiar with it; look it up)
TQ	(test questions)
1, 2, 3	(used to indicate a series of facts)
D	(shows disagreement with a statement or passage)
REF	(reference)
et al.	(and others)
bk	(book)
p	(page)
etc.	(and so forth)
V	(see)
VS	(see above)
SC	(namely)
SQ	(the following)
com.	(Commutative)
dis.	(Distributive)
APA	(associative property of addition)
AI	(additive inverse)
IPM	(identity property of multiplication)

When to Take Notes

To become a better note taker, you must know when to take notes and when not to take notes. The instructor will give cues that indicate which material is important. These cues include

- presenting facts or ideas
- writing on the board
- summarizing
- pausing
- repeating statements
- enumerating, such as "1, 2, 3" or "A, B, C"
- working several examples of the same type of problem on the chalkboard
- saying This is a tricky problem
- saying This is the most difficult step in the problem
- indicating that certain types of problems will be on the test, such as coin or age problems
- explaining boldface words.

You must learn the cues your instructor gives to indicate important material. If you are in doubt about the importance of the class material, do not hesitate to ask the instructor about its importance.

While taking notes, you may become confused about math material. At that point, take as many notes as possible, and do not give up on note taking. One thing you shouldn't do—bother the student next to you for an explanation. This can be annoying, and both of you will miss important information while you are talking.

As you take notes on confusing problem steps, leave extra space; then go back and fill in information that clarifies your misunderstanding of the steps in question. Ask your tutor or instructor for help with the uncompleted problem steps, and write down the reasons for each step in the space provided.

Another procedure for saving time while taking notes is to stop writing complete sentences. Write your main thoughts in phrases. Phrases are easier to jot down and easier to memorize.

SECTION 4.2   REVIEW

1. Note taking requires two basic strategies: _____ and
   _____.

2. List five abbreviations (and their meanings) that you use in your math class (one abbreviation and meaning per line).

   _____

   _____

   _____

   _____

   _____

3. List five cues that your instructor gives that indicate which material is important (one cue per line).

   _____

   _____

   _____

   _____

   _____

4.3 The Seven Steps in Math Note Taking

No matter what format you use to take notes, a good set of notes contains

- examples of problems
- explanations of how to complete the problems
- appropriate math vocabulary and rules that are used for each problem

In addition, it is wise to make your notes as organized as possible. It may get on your nerves for a while, but when you see that organized notes save time when you are studying, you will be more willing to take the time to organize them.

Many successful math students, whether in developmental math courses or calculus, use a form of the three-column note-taking system. It organizes the information, making it easy to review. While creating the three-column system, students are also learning the math, so it is not busywork. It takes a few weeks to feel comfortable with this system, but it is worth sticking with it. Many students have adopted this system and improved the efficiency of their study time.

The seven steps lead you through several important stages in the learning process. Steps 1 to 3 help you *record the information* accurately, completely, and in an organized fashion. Steps 4 to 7 help you *rehearse and understand the information*. These steps also help you learn to *recall the information*, thus helping you prepare for tests.

To prepare for taking notes this way, get a good chunk of paper ready. Here are some choices:

- Use regular paper, portrait or landscape style, and draw two lines to create three columns.
- See Figure 7, for a completed example.
- Some students purchase graph paper because it helps them keep numbers lined up correctly.
- Other students design the page on a computer and print out copies.
- Choose two columns next to each other to be the "Examples" and the "Explanations" columns.

FIGURE 7 Three-Column Note Taking System

Key Words / Rules	Example	Explanations
Solve a linear equation	$4(x + 4) + 3(x - 4) = 2(x - 1)$	Have to get x on one side of the = and numbers on the other side of the =.
Distributive Property	$4x + 16 + 3x - 12 = 2x - 2$	Multiply numbers to the left of the ( ) by each variable and number in the ( ).
Commutative Property	$4x + 3x + 16 - 12 = 2x - 2$	Regroup numbers and variables.
Combine like terms	$7x + 4 = 2x - 2$	
Additive Inverse Property	$7x - 2x + 4 = 2x - 2x - 2$ $5x + 4 = -2$	Subtract 2x from both sides in order to get variables all on the right side of the =.
Additive Inverse Property	$5x + 4 - 4 = -2 - 4$ $5x = -6$	Subtract 4 from both sides to get numbers all on the right side of the =.
Multiplicative Inverse Property	$\dfrac{5x}{5} = \dfrac{-6}{5}$	Divide both sides by 5 to get x by itself on the left side of the =.
Simplify	$x = -1\dfrac{1}{5}$	Write as a whole number and fraction.
	Insert new problem.	

- Label them as such. These are the two columns you will use all the time during class.
- Label the other column "Key Words/Rules." You might fill this column in during class but will probably complete it outside of class.

If you do this ahead of time, you do not have to draw lines frantically while the instructor is talking. If you forget or run out of paper, it might be quicker to fold the paper into three columns.

Draw a line dividing the end of one problem from the beginning of the next. Or, do one problem per page to leave plenty of room for notes. Graph paper can help you keep numbers and words lined up. Some students like to turn the note books or regular paper sideways and then organize their notes in three columns.

During class you will use two columns most of the time—the "Examples" and the "Explanations" columns. Sometimes, when instructors really want their students to use this method, they will organize their board work to reflect what the notes should look like. Most of the time, however, this does not happen.

The "Examples" column is where you will place the examples the instructor uses during class.

The "Explanations" column is where you will put the instructor's verbal explanations of the examples. If the instructor tends to explain by just doing the problem without talking about it, ask questions that will force the instructor to explain the problem verbally.

When the instructor simply explains a vocabulary word or rule, write the vocabulary word or a phrase naming the rule in the "Key Words/Rules" column. Write the explanation in the "Explanations" column. If the instructor provides an example, obviously it goes in the "Examples" column. If no example is provided, ask for one. Then if your wish is not granted, look one up in the textbook after class.

Here are the seven steps in math note taking. Remember that these are the basic steps. You can adapt them to work for you.

Step 1   *Record the instructor's example problems in the "Examples" column.*

Step 2   *Record the reasons for each step in the "Explanations" column. After completing a*

problem, draw a horizontal line, separating it from the next problem.

If you get lost during the lecture, place a question mark in this column and move on.

Step 3   *If possible, record the key words and rules in the first column.* Here is an opportunity for you to ask what rules are being used if the instructor is assuming that the class knows them and you do not. Otherwise, you will need to refer to the textbook, work with a tutor, or use online resources after class to complete this column.

It is important to understand the vocabulary words, rules, and concepts when learning how to complete a math problem. You will be able to figure out other problems that may be slightly different.

Step 4   *In order to keep everything in your head, use the notes to review.* Cover up the "Examples" and "Explanations" columns and quiz yourself on the vocabulary and rules. Or, fold back the "Key Words/Rules" column, look at the examples and explanations, and choose the correct key words and rules used to complete the examples.

Step 5   *Place a check mark by the key words and concepts that you did not know.*

Step 6   *Review the information that you checked until you understand and remember it.*

Step 7   *Develop a glossary of the math terms and rules that are difficult for you to remember.*

For a while you might spend more time organizing this system outside of class. Some students still have to depend on their current process, concentrating on getting the information down. Then they transfer the information into the three-column format. However, it is best to learn how to do most of the process during math class.

Remember the learning modalities and learning styles? This process can use all of them!

- *Kinesthetic:* Writing out the notes
- *Auditory:* Saying information out loud
- *Visual:* Looking at the notes and thinking; color-code each column with highlighters, making sure you do not smear the pencil writing

- *Group:* Working on the notes with one or two other people to make sure they are complete and accurate
- *Individual:* Reading the notes to yourself while waiting to meet someone

Current research is discovering that using language (vocabulary, discussion) to learn math helps students understand the concepts more deeply. Then completing many problems helps them master the skills and increase the speed that is important in preparing for tests. These notes will help you understand the math so that it will stick with you. You will remember more information. Homework will help you master the skills and increase the speed at which you can complete problems.

A Math Glossary

A math glossary is created to define math vocabulary in your own words. Since math is considered a foreign language, understanding the math vocabulary becomes the key to comprehending math. Creating a glossary for each chapter of your textbook will help you understand math.

Your glossary should include all words printed in boldface in the text and any words you do not understand form the lecture. If you cannot explain the math vocabulary in your own words, ask your instructor or tutor for help. You may want to use the last pages in your notebook to develop a math glossary for each chapter in your textbook. Some students use note cards. Review your math glossary before each test.

SECTION 4.3   REVIEW

1. List and describe the seven steps in math note taking.

    First Step: _____

    Second Step: _____

    Third Step: _____

    Fourth Step: _____

    Fifth Step: _____

    Sixth Step: _____

    Seventh Step: _____

2. Make several copies of the modified three-column note-taking system, and take notes using this system in your next math class. Or, develop the modified three-column note-taking system in your own notebook.

3. List two benefits of creating a math glossary.

    Benefit 1: _____

    Benefit 2: _____

4. From your current math chapter, list and define five words that you can put in your math glossary.

    Word 1: _____

    Word 2: _____

    Word 3: _____

Word 4: _____

Word 5: _____

4.4 Classroom Participation: Becoming an Active Questioner

Not that anyone should have to tell you this, but college isn't free. In fact, it is quite expensive. While you probably received a free education in high school, you are paying top dollar for your higher education, so you might as well take advantage of the tools at your disposal. If you have a question for your professor, you should never be afraid to ask it.

To begin with, there is no such thing as a stupid question. If you need something cleared up, even a basic idea, it is much better to have it explained by an expert than to have to figure it out on your own while doing your homework. Also, if you are confused about something, there is almost always another student in the class that will be stumped by the same information.

It is also true that professors generally prefer active questioners as these students help them stay on focus and to double-check that their message is being properly understood. It slows the professors down and gives them the peace of mind that they are doing a good job. Questions help them to further develop subtle points about complex processes, and are therefore an absolute must for a proper and successful lecture.

If you are afraid to ask questions on the spot, you should try reading parts of the text before class and walking in with a few questions written down in your notebook. Doing this should endear you to your professor, as he or she will see that you have an active interest in the subject matter and are therefore taking the course seriously.

In the case that you are absolutely too scared to ask a question in class, you can always email your professors—but this should be the absolute last case scenario, and should never become an ongoing habit.

We've touched on this a little bit in other areas of the text, but the importance of fully understanding the processes behind math problems cannot be understated. If you are confused about something, don't be afraid to ask the questions required to get you on the right track. Put that money of yours to work!

SECTION 4.4 REVIEW

1. Is there such thing as a stupid question? Why or why not?

   _____

   _____

   _____

2. Can asking questions actually help your professor teach? How so?

   _____

   _____

   _____

3. Name one of the methods mentioned that can help you walk in to class and ask questions with confidence.

   _____

   _____

   _____

4.5 | How to Rework Your Notes

The note-taking system does not stop when you leave the classroom. As soon as possible after class, rework your notes. You can rework the notes between classes or as soon as you get home. By reworking your notes as soon as possible, you can decrease the amount of information you forget. This is an excellent procedure for transferring math information from short-term memory to long-term memory.

REMEMBER

Most forgetting occurs right after the material is learned. You need to rework your notes as soon as possible. Waiting means that you probably will not understand what was written.

Here are six important steps in reworking your notes:

Step 1   *Rewrite the material you cannot read or will not be able to understand a few weeks later.* If you do not rework your notes, you will be frustrated when studying for a test if you come across notes you cannot read or completing your homework. Another benefit of rewriting the notes is that you immediately learn the new material. Waiting means it will take more time to learn the material.

Step 2   *Fill in the gaps.* Most of the time, when you are listening to a lecture, you cannot write down everything. It is almost impossible to write down everything, even if you know shorthand. Locate the portions of your notes that are incomplete. Fill in the concepts that were left out. In the future, skip two or three lines in your notebook page for anticipated lecture gaps.

Step 3   *Add additional key words and rules in the "Key Words/Rules" column.* These key words and rules are the ones that were not recorded during the lecture.

EXAMPLE

You did not know that you should add the *opposite* of 18 to solve a particular problem, and you incorrectly added 18. Put additional important key words and ideas (such as "opposite" and "negative of") in the notes; these are the words that will improve your understanding of math.

Step 4   *Make a problem log of those problems that the teacher worked in class.* The problem log is a separate section of your notebook that contains a listing of the problems (without explanations—just the problems) that your teacher worked in class. If your teacher chose those problems to work in class, you can bet that they are considered important. The problems in this log can be used as a practice test for the next exam. Your regular class notes will contain not only the solutions but also all the steps involved in arriving at those solutions and can be used as a reference when you take your practice test.

Write the problem on the left half of the page. Then save some room on the right side for writing out what the test instructions might be for the problem. Sometimes we forget what the test instructions are, and we miss points on the test by following the directions incorrectly. Skip several lines to complete the problem when you use the problem log as a practice test.

Step 5   *Make a calculator handbook and put in it your keystroke sequences.* The calculator handbook can be a spiral-bound set of note cards or a separate section of your notebook that holds only calculator-related information. Your handbook should also include an explanation of when that particular set of keystrokes is to be used.

Step 6 *Reflection and synthesis.* Once you have finished going over your notes, review the major points in your mind. Combine your new notes with your previous knowledge to gain a better understanding of what you have learned today.

If you are a group learner and even if you are not, it helps to review and reflect with other students.

Use a Tape Recorder

If you have a math class during which you cannot get all the information down while listening to the lecture, ask your instructor about using a tape recorder. To ensure success, the tape recorder must have a tape counter and must be voice activated.

The tape counter displays a number indicating the amount of tape to which you have listened. When you find you are in an area of confusing information, write the beginning and ending tape counter numbers in the left margin of your notes. When reviewing your notes, the tape count number will be a reference point for obtaining information to work the problem. You can also reduce the time it takes to listen to the tape by using the pause button to stop the recording of unnecessary material.

Ask Questions

To obtain the most from a lecture, you must ask questions in class. By asking questions, you improve your understanding of the material and decrease your homework time. By not asking questions, you create for yourself unnecessary confusion during the remainder of the class period. Also, it is much easier to ask questions in class about potential homework problems than it is to spend hours trying to figure out the problems on your own at a later time.

If you are shy about asking questions in class, write down the questions and read them to your instructor. If the instructor seems confused about the questions, tell him or her that you will discuss the problem after class. To encourage yourself to ask questions, remember the following:

- You have paid for the instructor's help.
- Five other students probably have the same question.
- The instructor needs feedback on his or her teaching to help the class learn the material.
- There is no such thing as a stupid question
- If you have trouble asking questions, then email your instructor the question.

Record Each Problem Step

The final suggestion on note taking is to record each step of every problem written or verbally explained. By recording each problem step, you begin overlearning how to work the problems. This will increase your problem-solving speed during future tests. If you get stuck on the homework, you will also have complete examples to review.

The major reason for recording every step of a problem is to understand how to do the problem while the instructor is explaining it instead of trying to remember unwritten steps later. Recording each step will help when completing your homework because each step of a similar problem can be reviewed to help you solve the homework problem. Although it may seem time consuming, it pays off during homework and test time.

SECTION 4.5   REVIEW

1. List and describe the six steps in reworking your notes.

First Step: _____

Second Step: _____

Third Step: _____

Fourth Step: _____

Fifth Step: _____

Sixth Step: _____

2. How can using a tape recorder in class improve your learning?

_____

3. List three reasons for asking questions in class.

First Reason: _____

Second Reason: _____

Third Reason: _____

4. What is the major reason for recording each problem step?

_____

CHAPTER 4   REVIEW

1. Sitting in the golden triangle of success can help you pay more _____ during class. (Answer on page 52)

2. Class time is an intense _____ that should not be wasted. (Answer on page 53)

3. Auditory learners should review the math videotape and _____ back the important information.

4. The goal of note taking is to take the _____ amount of notes and get the _____ amount of information.

5. Using abbreviations can help you write _____ and listen _____.

6. Your _____ should include any words you don't know and their explanations.

7. Rework your notes to transfer information from _____-_____ memory to _____-_____ memory.

8. A _____ _____ is a separate section in your notebook that contains problems that your teacher worked in class.

9. Asking _____ in class will improve your understanding and decrease your home-work time.

10. _____ each step of the problem will increase your problem-solving speed and memory.

What is the most important information you learned from this chapter?

_____
_____
_____
_____

How can you immediately use it?

_____
_____
_____
_____

DAN'S TAKE REVIEW

1. What were the consequences of Dan not paying attention while he was trying to take notes?

_____
_____
_____

2. Were his distractions limited to new technologies, or has there always been a certain level of distraction that might vary depending upon a given situation?

    _____

    _____

    _____

3. Having read what you have in this chapter, what would you tell him to do to avoid distractions in the future?

    _____

    _____

    _____

How to Improve Your Reading, Homework, and Study Techniques

CHAPTER 5

Reading a math textbook is more difficult than reading other textbooks. Math textbooks are written differently than English and social science textbooks. Math textbooks contain condensed material and, therefore, take longer to read.

Mathematicians can reduce a page of writing to one paragraph using math formulas and symbols. To make sure you understood that same information, an English instructor would take that original page of writing and expand it into two pages. Mathematicians pride themselves on how little they can write and still cover the concept. This is one reason it may take you two to three times as long to read your math text as it would any other text.

REMEMBER

Reading your math text will take longer than reading your other texts.

Math students are expected to know how to do their homework; however, most math students do not have a homework system. Most students begin their homework by going directly to the problems and trying to work them. When they get stuck, they usually quit. This is not a good homework system.

A good homework system will improve your homework success and math learning at the same time.

DAN'S TAKE

College students are busy people. With hours of work and multiple deadlines, procrastination seems very appealing. Why suffer every night when you can put things off?

This type of logic stayed with me most of the way through college, and to be honest, it worked fairly well. Then I took my first college math course.

In this class, homework was going to be collected only four times during the semester. I thought I would need to spend only four nights the entire semester doing math homework. Granted, I knew that these nights would be rough on me and require hours of uninterrupted study, but to limit myself to just a few of them over the course of the semester seemed smarter to me than stressing out every single day.

I went to class every day and took very good notes. Then I went home at night and let them gather dust. By the time the first test rolled around, I had not done a single math problem. The night before the exam I sat down, opened my book, and looked at the homework assignment list. There were *ten chapters* due! I worked the whole night, and got a low C on my first test.

When I actually thought about it, I got really mad at myself. The homework was the easiest way to earn points. All I had to do was complete the work and I would have been given full credit. I also didn't realize that taking good notes was not enough to succeed in a college-level math course. By putting the homework off, I forgot how to do the math and had to relearn it all in one night.

Math is totally different from other courses such as science, history, and literature. Putting off homework can cause you to fall far behind, and catching up is very difficult. It's much easier to do a little bit every day than to try to cram an entire semester's worth of material into one night!

5.1   How to Read a Math Textbook

The way you read a math textbook is different from the traditional way students are taught to read textbooks in high school and college. Students are taught to read quickly or skim the material. If you do not understand a word, you are supposed to keep on reading. Instructors of other courses want students to continue to read so they can pick up the unknown words and their meanings from context.

This reading technique may work with your other classes, but using it in your math course will be totally confusing. If you skip some major concept words or boldface words, you will not understand the math textbook or be able to do the homework. Reading a math textbook takes more time and concentration than reading your other textbooks.

If you have a reading problem, it would be wise to take a developmental reading course before taking math. This is especially true with reform math courses, where reading and writing are emphasized.

Reform math classes deal more with word problems than traditional math courses do. If you cannot take the developmental reading course before taking math, then take it during the same semester as the math course.

Ten Steps in Understanding Reading Materials

There are several appropriate steps in reading a math textbook:

Step 1 *Skim the assigned reading material.* Skim the material to get a general idea about the major topics. Read the chapter introduction and each section summary. You do not want to learn the material at this time; you simply want to get an overview of the assignment. Then think about similar math topics that you already know.

> **EXAMPLE**
>
> Skimming will allow you to see if problems presented in one chapter section are further explained in later chapter sections.

Step 2 *As you skim the chapter, circle (using a pencil) the new words that you do not understand.* If you do not understand these new words after reading the assignment, ask the instructor for help. Skimming the reading assignments should take only 5 to 10 minutes.

Step 3 *Put all your concentration into reading.* While reading the textbook, highlight the material that is important to you. However, do not highlight more than 50% of any page, because doing so will not sufficiently narrow the material for future study. Especially highlight the material that is also discussed in the lecture. Material discussed in both the textbook and the lecture will usually appear on the test. The purpose of highlighting is to emphasize the important material for future study. Do not skip reading assignments.

> **REMEMBER**
>
> Reading a math textbook is very difficult. It might take you half an hour to read and understand just one page.

Step 4 *When you get to the examples, go through each step.* If the example skips any steps, make sure you write down each one of those skipped steps in the textbook for better understanding. Later on, when you go back and review, the steps will already be filled in. You will understand how each step was completed. Also, by filling in the extra steps, you will start to overlearn the material for better recall on future tests.

Step 5 *Mark the concepts and words that you do not know.* Maybe you marked them the first time while skimming. If you understand them now, erase the marks. If you do not understand the words or concepts, then reread the page or look them up in the glossary. Try not to read any further until you understand all the words and concepts.

Step 6 If you do not clearly understand some words or concepts, add these words to the note-taking glossary in the back of your notebook. Your glossary will contain the boldface words that you do not understand. If you have difficulty understanding the boldface words, ask the instructor for a better explanation. You should know all the words and concepts in your notebook's glossary before taking the test.

Step 7 *If you do not understand the material, follow these eight points, one after the other, until you do understand the material:*

Point 1—Go back to the previous page and reread the information to maintain a train of thought.

Point 2—Read ahead to the next page to discover if any additional information better explains the misunderstood material.

Point 3—Locate and review any diagrams, examples, or rules that explain the misunderstood material.

Point 4—Read the misunderstood paragraph(s) several times aloud to better understand their meaning.

Point 5—Refer to your math notes for a better explanation of the misunderstood material.

Point 6—Refer to another math textbook, computer software program, or videotape that expands the explanation of the misunderstood material.

Point 7—Define exactly what you do not understand and call your study buddy for help.

Point 8—Contact your math tutor or math instructor for help in understanding the material.

Step 8 *Recall important material.* Try to recall some of the important concepts and rules. If you have difficulty recalling this information, look at the highlighted areas of the text, boldface words, or formulas. If you are a visual learner, write down these words several times on a separate sheet of paper. If you are an auditory learner, repeat these words several times out loud. If you are a kinesthetic (hands-on) learner, write out the formulas twice on two separate cards. Cut one of them up into puzzle pieces. Mix up the parts of the puzzle. Turn over the other formula card so you cannot see it. Put the puzzle back together. Do this several times. For boldface words, write the word on one card and its explanation twice on two other separate cards. Make another puzzle. Mix up the pieces of the explanation and put them back together. Then add these words and formulas to your note-taking glossary. Recalling important material right after you read it improves your ability to remember that information.

Step 9 *Reflect on what you have read.* Combine what you already know with the new information that you just read. Think about how this new information enhances your math knowledge. Prepare questions for your instructor on the confusing information.

Ask those questions at the next class meeting.

Step 10 *Write anticipated test questions.* Research has noted that students have about 80% accuracy in predicting test questions. Think about what is the most important concept you just read and what problems the instructor could give you that would test the knowledge of that concept. Make up three to five problems and add them to your problem log (this was part of your note-taking system). Indicate in the log that these are your questions, not the instructor's.

By using this reading technique, you narrow down the important material to be learned. You skim the textbook to get an overview of the assignment. You carefully read the material and highlight the important parts. You then add unfamiliar words or concepts to your note-taking glossary.

R E M E M B E R
You should review the highlighted material before doing the homework problems, and you should learn 100% of the glossary before taking the test.

How Reading Ahead Can Help

Reading ahead is another way to improve learning. If you read ahead, do not expect to understand everything. Read ahead two or three sections and put question marks (in pencil) by the material you do not understand.

When the instructor starts discussing that material, have your questions prepared and take good notes. Also, if the lecture is about to end, ask the instructor to explain the confusing material in the textbook. Reading ahead will take more time and effort, but it will better prepare you for the lectures.

How to Establish Study Period Goals

Before beginning your homework, establish goals for the study period. Do not just do the homework problems.

Ask yourself this question: "What am I going to do tonight to become more successful in math?"

By setting up short-term homework goals and reaching them, you will feel more confident about math. This also improves your self-esteem and helps you become a more internally motivated student. Set up homework tasks that you can complete. Be realistic.

Study period goals are set up either on a time-line basis or an item-line basis. Studying on a time-line basis is studying math for a certain amount of time.

EXAMPLE

You may want to study math for an hour and then switch to another subject. You will be studying on a time-line basis.

Studying on an item-line basis means you will study your math until you have completed a certain number of homework problems.

EXAMPLE

You might set a goal to study math until you have completed all the odd-numbered problems in the chapter review. The odd-numbered problems are the most important problems to work, because, in most texts, they are answered in the answer section in the back of the book. Such problems provide the opportunity to recheck your work if you do not get the answer correct. Once you have completed these problems, do the even-numbered problems.

No matter what homework system you use, remember this important rule: Always finish a homework session by understanding a concept or doing a homework problem correctly.

Do not end a homework session with a problem you cannot complete. You will lose confidence, since all you will think about is the last problem you could not solve instead of the 50 problems you solved correctly. If you do quit on a problem you cannot solve, rework some of the problems you have done correctly.

 REMEMBER

Do not end your study period with a problem you cannot complete.

SECTION 5.1 REVIEW

1. List and describe the 10 steps in reading a math textbook.

Step 1: _____

Step 2: _____

Step 3: _____

Step 4: _____

Step 5: _____

Step 6: _____

Step 7: _____

Step 8: _____

Step 9: _____

Step 10: _____

2. After trying each of the 10 steps, make up your own condensed version of these steps.

Step 1: _____

Step 2: _____

Step 3: _____

Step 4: _____

Step 5: _____

Step 6: _____

Step 7: _____

Step 8: _____

Step 9: _____

Step 10: _____

3. List two reasons why reading ahead can improve your learning.

First Reason: _____

Second Reason: _____

4. Describe the two types of study period goals. Which one do you use?

First Type: _____

_____

Second Type: _____

_____

5.2   How to Do Your Homework

Doing your homework can be frustrating or rewarding. Most students jump right into their homework, become frustrated, and stop studying. These students usually go directly to the math problems and start working them without any preparation. When they get stuck on one problem, they flip to the back of the text for the answer. Then they either try to work the problem backward to understand the problem steps or they just copy down the answer.

Other students go to the solution guide or the textbook website and copy the steps. After getting stuck several times, these students will inevitably quit doing their homework. Their homework becomes a frustrating experience and they may even quit doing their math homework altogether.

Motivation decreases every time a student does not feel that they learned anything from his or her homework to the point that they may begin skipping classes and subsequently fail. Students may not have to successfully complete each and every problem, but they do need to feel successful. In addition, they must also develop a plan on how to get help solving the problems they cannot complete.

The Importance of Doing Your Homework

Do you know the reasons for doing math homework? I have asked this question to math instructors and to tutors and was surprised by their answers.

Students have told me that homework is assigned to waste their time because the instructor did not want to teach them or because the instructor did not like them and wanted them to suffer. Reinforcing this theory, the instructors did not even grade the homework or return it to them in time to prepare for the test. Even in this instance, the student's perception is not necessarily true. Instructors almost always want you to learn and will assign you homework only in order to improve your learning and success.

Now let's look at the importance of doing your homework. In sports or in learning to play a musical instrument, you need to practice in order to have a good game or perform well in a recital. In the case of math, homework is that "practice" so you can take the information taught in the classroom and put it into abstract and long-term memory (you will learn more about this concept in Chapter 6). You need to be able to practice the math problems often enough to understand the concepts and then automatically recognize the problems and know how to solve them. In other words, you do your homework to remember how to solve the problems during the test. However, just memorizing how to do specific problems is not enough. Most instructors will not put homework problems on the test, so it becomes important to have a full grasp on the concepts behind them in order for the problems to become interchangeable.

Doing your homework provides an excellent opportunity to practice the problems as they might appear on a test and thereby reinforce the mathematical concepts used to solve them. This is perhaps the best way to figure out what problems you cannot solve *before* taking the test. Not doing your homework or not completing your homework means that the information never reaches long-term memory. This leads to poor math learning and potential math failure. Successfully completing homework leads to a positive learning experience.

Now that you know its importance, has anyone every taught you the best way to complete your math homework? Probably not. The next section will teach you a math homework system that has helped thousands of students better learn math and thereby improve their grades.

Ten Steps in Doing Your Homework

To improve your homework success and learning, refer to these 10 steps:

Step 1 *Review the textbook material that relates to the homework.* A proper review will increase your chances of successfully completing your homework. If you get stuck on a problem, you will have a better chance of finding similar problems. If you do not review prior to doing your homework, you could get stuck and not know where to find help in the textbook.

REMEMBER
To be successful in learning the material and in completing homework assignments, you must first review your textbook.

Step 2 *Review your lecture notes that relate to the homework.* If you cannot understand the explanation in the textbook on how to complete the homework assignment, then review your notes.

REMEMBER
Reviewing your notes will give you a better idea about how to complete your homework assignment.

Step 3 *Do your homework as neatly as possible.* Doing your homework neatly has several benefits. When approaching your instructor about problems with your homework, he or she will be able to understand your previous attempts to solve the problem. The instructor will easily locate the mistakes and show you how to correct the steps without having to decipher your handwriting. Another benefit is that when you review for midterm or final exams, you can quickly relearn the homework material without having to decipher your own writing.

REMEMBER

Neatly prepared homework can help you now and in the future.

Step 4   *When doing your homework, write down every step of the problem.* Even if you can do the step in your head, write it down anyway. This will increase the amount of homework time, but you will be over-learning how to solve problems, which improves your memory. Writing down every step is an easy way to memorize and understand the material. Another advantage is that when you rework the problems you did wrong, it is easy to review each step and to find your mistake.

REMEMBER

In the long run, writing down every step of your homework problems will save you time and frustration.

Step 5   *Understand the reasons for each problem step and check your answers.* Do not get into the bad habit of memorizing how to do problems without knowing the reasons for each step. Many students are smart enough to memorize the procedures required to complete a set of homework problems. However, when similar homework problems are presented on a test, some of these students cannot solve them. To avoid this dilemma, keep reminding yourself about the rules, laws, or properties used to solve problems.

EXAMPLE

Problem: $2(a + 5) = 0$. What property allows you to change this equation to $2a + 10 = 0$? *Answer:* the distributive property.

Once you know the correct reason for going from one step to another in solving a math problem, you can solve any problem requiring that property.

Students who simply memorize how to do problems instead of understanding the reasons for correctly working the steps are more likely to fail their math course.

Checking your homework answers should be part of your homework duties. Checking your answers can improve your learning and help you prepare for tests.

Check the answers of the problems for which solutions are not given. These may be the even-numbered or odd-numbered problems or the problems not answered in the solutions manual.

First, check each answer by estimating the correct answer.

EXAMPLE

If you are multiplying 2.234 by 5.102, remember that 2 times 5 is 10. The answer should be a little over 10.

You can also check your answers by substituting the answer back into the equation or doing the opposite function required to answer the question. The more answers you check, the faster you will become. This is very important because increasing your answer-checking speed can help you catch more careless errors on future tests.

Step 6   If you do not understand how to do a problem, refer to the following points:

Point 1—Review the textbook material that relates to the problem.

Point 2—Review the lecture notes that relate to the problem.

Point 3—Review any similar problems, diagrams, examples, or rules that explain the misunderstood material.

Point 4—Refer to another math textbook, solutions guide, math computer program software, or videotape to obtain a better understanding of the material.

Point 5—Call your study buddy.

Point 6—Skip the problem and ask your tutor or math instructor for help as soon as possible.

Step 7   *Always finish your homework by successfully completing problems.* Even if you get stuck, go back and successfully complete previous problems before quitting. You want to end your homework assignment with feelings of success.

Step 8 *After finishing your homework assignment, recall to yourself or write down the most important concepts you have learned.* Recalling this information will increase your ability to learn these new concepts. Additional information about Step 8 will be presented later in this chapter.

Step 9 *Make up note cards containing hard-to-remember problems or concepts.* Note cards are an excellent way to review material for a test. More information on the use of note cards as learning tools is presented later in this chapter.

Step 10 *Do not fall behind.* As mentioned in Chapter 1, learning math is a sequential process. If you get behind, it is difficult to catch up because each topic builds on the previous topic. Falling behind in math is like going to Spanish class without learning the last set of vocabulary words. The teacher will be talking to you using the new vocabulary, but you will not understand what is being said.

Do Not Fall Behind

To keep up with your homework, it is recommended that you work on it every school day and even on weekends. Working on your homework for one-half hour each day for two days in a row is better than working on it for one hour every other day.

If you have to get behind in one of your courses, *make sure it is not math.* Fall behind in a course that does not require a sequential learning process, such as psychology or history. After using the ten steps in doing your homework, you may be able to combine two steps into one. Find your best combination of homework steps and use them.

REMEMBER
Getting behind in math homework is the fastest way to fail the course.

SECTION 5.2 REVIEW

1. List the 10 steps you should follow in doing your math homework.

    Step 1: _____

    Step 2: _____

    Step 3: _____

    Step 4: _____

    Step 5: _____

    Step 6: _____

    Step 7: _____

    Step 8: _____

    Step 9: _____

    Step 10: _____

2. After trying each of the 10 steps, make up your own condensed version of these steps.

    Step 1: _____

    Step 2: _____

Step 3: _____

Step 4: _____

Step 5: _____

Step 6: _____

Step 7: _____

Step 8: _____

Step 9: _____

Step 10: _____

3. How will reviewing your notes or your math text help you complete your homework more successfully?

_____

_____

_____

4. List two ways you can check your homework answers.

First Way: _____

Second Way: _____

5. What happens to students who fall behind in their math homework?

_____

_____

_____

5.3   How to Solve Word Problems

The most difficult homework assignment for most math students is working story/word problems. Solving word problems requires excellent reading comprehension and translating skills.

Students often have difficulty substituting algebraic symbols and equations for English terms. But once an equation is written, it is usually easily solved. To help you solve word problems, follow these ten steps:

Step 1   *Read the problem three times.* Read the problem quickly the first time as a scanning procedure. As you are reading the problem the second time, answer these three questions:

1. *What is this problem asking me?* (This is usually at the end of the problem.)
2. *What is this problem telling me that is useful?* (Cross out unneeded information.)
3. *What is this problem implying?* (This is usually something you have been told to remember.) Now read the problem a third time to make sure that you fully understand its meaning.

Step 2   *Draw a simple picture of the problem to make it more real to you* (e.g., a circle with an arrow can represent travel in any form— by train, by boat, by plane, by car, or by foot).

Step 3   *Make a table of information and leave a blank space for the information you are not told.*

Step 4   *Use as few unknowns in your table as possible.* If you can represent all the unknown information in terms of a single letter, do so! When using more than one unknown, use letters that remind you of each unknown. Then write down what each unknown represents. This eliminates the problem of assigning the right answer to the wrong unknown. Remember that you have to create as many separate equations as you have unknowns.

Step 5   Translate the word problem into an algebraic equation using the lists of terms in Figures 8 and 9. Remember, English terms are sometimes stated in a different order than the corresponding algebraic terms.

Step 6   *Immediately retranslate the equation, as you now have it, back into words.* The translation may not sound like a normal word problem, but the meaning should be the same as the original problem. If the meaning is not the same, the equation is incorrect and needs to be rewritten. Rewrite the equation until it means the same as the word problem.

Step 7   *Review the equation to see if it is similar to equations from your homework and if it makes sense.* Some formulas dealing with specific word problems may need to be rewritten. Distance problems, for example, may need to be written solving for each of the other variables in the formula. Given distance = rate × time, time = distance/rate, and rate = distance/time. Usually, a distance problem will identify the specific variable for which you are to solve.

Step 8   *Solve the equation using the rules of algebra.*

REMEMBER

Whatever is done to one side of the equation must be done to the other side of the equation. The unknown must end up on one side of the equation, by itself. If you have more than one unknown, then use the substitution method or the elimination method to solve the equations.

Step 9   *Look at your answer to see if it makes sense.*

EXAMPLES

If tax is added to an item, it should cost more; if a discount is applied to an item, it should cost less. Is there more than one answer? Does your answer match the original question? Does your answer have the correct units?

Step 10   *Substitute your answer back into the original equation to see if it is correct.* If one side of the equation equals the other side of the equation, then you have the correct answer. If you do not have the correct answer, go back to Step 5.

The most difficult part of solving word problems is translating part of a sentence into algebraic symbols and then into algebraic expressions. Review Figure 8 (Translating English Words and Phrases into Algebraic Symbols) and Figure 9 (Translating English Phrases into Algebraic Expressions).

How to Recall What You Have Learned

After completing your homework problems, a good visual learning technique is to make note cards. Note cards are 3" × 5" index cards on which you place information that is difficult to learn or material you think will be on a test.

On the front of each note card, write a math problem or information that you need to know. Color-code the important information in red or blue. On the back of the note card, write how to work the problem or explain important information.

EXAMPLE

If you are having difficulty remembering the rules for multiplying positive and negative numbers, write some examples on the front of a note card and write the answers on the back.

Make note cards on important information you might forget. Every time you have 5 spare minutes, pull out your note cards and review them. You can glance at the front of the card, repeat the answer to yourself, and check it by looking at the back of the card. If you are correct and know the information

FIGURE 8   Translating English Words and Phrases into Algebraic Symbols

English Word or Phrase	Algebraic Symbol
Sum	+
Add	+
In addition	+
More than	+
Increased	+
In excess	+
Greater	+
Decreased by	−
Less than	−
Subtract	−
Difference	−
Diminished	−
Reduced	−
Remainder	+
Times as much	×
Percent of	×
Product	×
Interest on	×
Per	÷
Divide	÷
Quotient	÷
Quantity	( )
Is	=
Was	=
Equal	=
Will be	=
Results	=
Greater than	>
Greater than or equal to	≥
Less than	<
Less than or equal to	≤

FIGURE 9   Translating English Phrases into Algebraic Expressions

English Phrases	Algebraic Expression
Ten more than x	$x + 10$
A number added to 5	$5 + x$
A number increased by 13	$x + 13$
Five less than 10	$10 - 5$
A number decreased by 7	$x - 7$
Difference between x and 3	$x - 3$
Difference between 3 and x	$3 - x$
Twice a number	$2x$
Ten percent of x	$0.10x$
Ten times x	$10x$
Quotient of x and 3	$x/3$
Quotient of 3 and x	$3/x$
Five is 3 more than a number	$5 = x + 3$
The product of 2 and a number is 10	$2x = 10$
One-half of a number is 10	$x/2 = 10$
Five times the sum of x and 2	$5(x + 2)$
Seven is greater than x	$7 > x$
Five times the difference of a number and 4	$5(x - 4)$
Ten subtracted from 10 times a number is that number plus 5	$10x - 10 = x + 5$
The sum of 5x and 10 is equal to the product of x and 15	$5x + 10 = 15x$
The sum of two consecutive integers	$(x) + (x + 1)$
The sum of two consecutive even integers	$(x) + (x + 2)$
The sum of two consecutive odd integers	$(x) + (x + 2)$

on the card, do not put it back in the deck. Mix up the cards you do not know and pick another card. Keep doing this until there are no cards left in the deck.

If you are an auditory learner, use a tape recorder instead of note cards. Record the important information as you would on the front of a note card. Then leave a blank space on the recording before recording the answer. Play the tape back. When you hear the silence, put the tape on pause. Then say the answer out loud to yourself. Take the tape player off pause and see if you were correct.

Review What You Have Learned

After finishing your homework, close the textbook and try to remember what you have learned. Ask yourself these questions: What major concepts did I learn tonight? What test questions might the instructor ask on this material?

For about 3 to 4 minutes, recall the major points of the assignment, especially the areas you had difficulty understanding. Write down questions for the instructor or tutor. Since most forgetting occurs right after the material is learned, this short review will help you retain the new material.

SECTION 5.3 REVIEW

1. List the 10 steps in solving word problems.

   Step 1: _____

   Step 2: _____

   Step 3: _____

   Step 4: _____

   Step 5: _____

   Step 6: _____

   Step 7: _____

   Step 8: _____

   Step 9: _____

   Step 10: _____

2. What is your best way to recall what you have learned?

   _____

   _____

   _____

3. How does reviewing right after doing your homework help you remember?

   _____

   _____

   _____

5.4 How to Work with a Study Buddy

You need to have a study buddy when you miss class or when you do your homework. A study buddy is a friend or classmate who is taking the same course. You can find a study buddy by talking to your classmates or making friends in the math lab. Try to find a study buddy who knows more about math than you do. Tell the class instructor that you are trying to find a study buddy and ask which students make the best grades.

Meet with your study buddy several times a week to work on problems and to discuss math. If you miss class, get the notes from your study buddy so you will not fall behind.

Call your study buddy when you get stuck on your homework. You can solve math problems over the phone. Do not sit for half an hour or an hour trying to work one problem; that will destroy your confidence, waste valuable time, and possibly alienate your study buddy. Think how much you could have learned by trying the problem for 15 minutes and then calling your study buddy for help. Spend, at the maximum, 15 minutes on one problem before going on to the next problem or calling your study buddy.

REMEMBER

A study buddy can improve your learning while helping you complete your homework assignments. Just do not overuse your study buddy or expect your study buddy to do your homework for you.

SECTION 5.4   REVIEW

1. How do you select a study buddy? _____

2. How can a study buddy help you learn math? _____

3. Who is your study buddy? _____

5.5 | The Benefits of Study Breaks

Psychologists have discovered that learning decreases if you do not take study breaks. Therefore, take a break after studying math for 45 minutes to 1 hour.

If you have studied for only 15 or 20 minutes and feel you are not retaining the information or your mind is wandering, take a break. If you continue to force yourself to study, you will not learn the material. After taking a break, return to studying.

If you still cannot study after taking a break, review your purpose for studying and your educational goals. Think about what is required to graduate. It will probably come down to the fact that you will have to pass math. Think about how studying math today will help you pass the next test; this will increase your chances of passing the course and of graduating.

On an index card write three positive statements about yourself and three positive statements about studying. Look at this index card every time you have a study problem. Use every opportunity available to reinforce your study habits.

SECTION 5.5   REVIEW

1. How can study breaks improve your learning?

_____

_____

_____

_____

2. What do you do when you don't want to study? How do you get started?

_____

_____

_____

_____

5.6 Using Online/Computer Resources to Support Learning

Many math textbooks include CDs and access to online websites that provide further practice and learning. There are even online homework programs that offer instant feedback and help. While these resources may seem unfamiliar or difficult to use at first, online support and CDs can improve some student learning and are worth a try.

What Are Online/Computer Resources?

Computer and web resources come in a variety of formats. Some math texts have CDs that can provide additional practice exercises and suggestions to improve learning. Other books come with passwords to websites that contain online tutorials and extra practice problems. Other publishers provide live tutorial help, such as *Enhanced WebAssign®*, which is offered by Cengage and provides 24-hour-a-day support for students. There are even math tutorial sites that are not meant for use with your textbook—just Google "math" and see what comes up! Finally, your own math department, math lab, or learning resource center may have other CDs, computer programs, or websites that can help you. Find out what online/computer resources are available to supplement your math learning.

Computer Technology Requirements

It is important to find out if the computer you use on a regular basis will support the online/computer program resources for further math learning. Check the operating or installation instructions for any program or website you want to use to make sure that your computer meets the minimum hardware and Internet connection requirements.

If your own computer is inadequate, do not despair and decide not to try the online or CD resources. Most campuses provide excellent computer and online access for students, but remember that most computers in the lab may not let you save your work on their hard drives. Be sure to bring your own flash drive or recordable CD so you can save your work.

After finding a computer that is adequate for the program's technical requirements, do a trial run. It is helpful if you can do this in a lab or library that has technical assistants in case you run into problems during installation or setup. Once the math program or website comes up, explore the different aspects of that program. If you have difficulty running the programs or need input on which program may be best for you, don't hesitate to ask your instructor for help. If your instructor is not available, then ask the lab learning assistants or a tutor. If they are not available, then ask your fellow students for help.

During the Semester

Using online/computer resources can help in various ways. First, try each online/computer resource available to you and see which ones work best for you. Some students like the drill and practice that is provided by an online quiz, while other students want more in-depth knowledge on the reason for doing each step and prefer interactive tutorials. Second, try using the online/computer resources both before and after your class or tutoring session to see which schedule helps you most. Some students like to use these resources as a preview to a math lesson or as a review after the lesson. Sometimes it might help to use the resources before *and* after the lesson.

Treat the time you spend using the resources as if it were class or lab time. First, set up a time and place to use the online/computer resources. Try to use the resources at the same place and same time. For example, if you have an hour between classes, go to the lab and practice, or practice right when you get home. This is a good way to develop excellent study habits. Second, keep a log for several weeks on which resources you used and how much time you spent on them. If it is obvious that certain times, places, or programs work best for you, keep the same study system. If your system is not working for you, adjust and revise it. Ask a lab assistant or instructor for advice.

OVERCOMING TECHNOPHOBIA

If you are experiencing technophobia (fear of technology, including computers), this is a wonderful time to overcome it. You can start with these strategies:

- Get in touch with the benefits of technology. Being comfortable with computers can give you an edge in almost every aspect of being a student, from doing library research to planning your career. In the eyes of many employers, experience with computers is sometimes a necessity and almost always a plus.
- Sign up for a computer class for beginners.
- Ask questions. When it comes to computers, there truly aren't any "dumb" questions.
- Find a competent teacher—someone who remembers what it was like to know nothing about computers.
- Just experiment. Sit down, do something, and see what happens. Short of dropping a computer or hitting it with a hammer, you can't hurt the thing.
- Remember that computers are not always user friendly—at least not yet. Learning how to use them takes patience and time. Knowing this up front can put you at ease and prepare you for the cyberspace adventures ahead.
- Also remind yourself of past successes in making transitions. As far as technology is concerned, this includes everything from writing with a pen to driving a car. You're already mastering a major life change—the transition to higher education. In doing so, you've shown that you have what it takes to tame any technology that enters your life.

Learning Styles and Online/Computer Learning

Your learning modality style may determine how helpful online/computer learning may be for you. Visual learners who like to learn by themselves usually work well with these types of resources, while auditory learners who need to hear the information may not like the resources unless they are also auditorily enhanced. Also, hands-on learners may not benefit from these programs unless there is some type of movement with the learning. Nevertheless, these programs are worth a try no matter what the learning style. Why? We all learn visually to some extent, and auditory and hands-on learners can adjust these resources to be more auditory or hands-on. For instance, students can read the screens out loud as they are using them, or can work with partners. Learning styles are just preferred ways to learn. Productive students learn how to use all available modalities when learning math.

Evaluate the Experience

After trying these programs for several weeks, evaluate whether they are helping you improve your learning or your grades. Evaluate each program. Take into consideration how much time you spent with each program. Look through your log to remind yourself of these details. You may have to wait until the next major test to see if your grades improve. Ask yourself, Did the resources improve my learning? and Did the resources improve my test scores? If the answer to either of these questions is yes, then keep on using them!

SECTION 5.6   REVIEW

1. Look in the preface of your math textbook and list some of the online/computer resources that are included with your book.

_____

_____

_____

2. Which resource listed sounds as if it would be the most helpful to you? Why?

_____

_____

_____

3. Remember your learning modality style from Chapter 2. How could you use one of the online/computer resources so that it worked with your personal learning modality?

_____

_____

5.7   The Benefits of Time Management and Study Breaks

Now that you have learned some of the best ways to read the textbook, complete your homework, solve word problems, work with a study buddy, and use online resources, it is important to figure out when will you have time to complete these activities. Many students have told me that they don't even have enough time to do their homework, let alone all these other activities that **will** improve their grades. **I do not believe them.** If you are truly an adult, then you can find ways to control your time. The truth is, many students simply choose not to control their time and, therefore, allow time to control them. For many students it comes down to priorities and making good choices.

If you want to obtain a better career for yourself or your family, then making time to study math should be your top priority. Most math students need to study between 8 and 10 hours a week to get an A or a B. Notice that I did not mention Cs because students who make a C in an algebra-based course will most likely fail their next algebra course.

Another way to improve math grades is to become an effective math learner. The truth is, throughout your academic careers, most of you have probably never been taught how to learn math. This, of course, is what this particular text is all about. Some general learning efficiency suggestions are to

- Set up a study time to review your notes or do your homework as soon as possible after math class.

- Leave a 1-hour opening in your schedule right after math class.
- Go to the math lab or Learning Resource Center right after math class and/or right before math class.
- Study math right when you get home, when you are alert.
- Review your problems right before you go to bed the night before your test (without watching TV or playing video games).
- Find the best time to study math according to your personal biological clock (morning, noon, early afternoon, night).
- Use the other suggestion in this text to improve your learning.

The next step is to list your priorities, as well as how much time you need to spend on your top priorities. Examples of top priorities may include work, attending class, family, sleep, eating, church, or others. You need to rank these priorities from most important to the least important. Then, next to each activity, put the number of hours you can allow for that priority. Remember you need to study math about 10 hours a week to get an A or a B. Also, it is important to remember that there are 168 hours in a week.

Next, you need to write down the times you have decided to study math. You may want to put the times in your notebook or on a 3" × 5" card and then put the card on the refrigerator. You will then need to remind yourself about the study times and keep a weekly log of the hours you have actually studied.

Next you need to set a goal for your math grade. After studying for several weeks and receiving your first major math test grade, you may have to adjust your study hours based upon the results.

Of all the students with whom I have worked who have flunked out of college or high school, most pointed to time management as one of the major culprits. Proper time management can help you become successful in college, just as bad time management can cause you to fail a math course and eventually drop out. The choice is yours!

SECTION 5.7   REVIEW

1. How can time management help you or hurt you while trying to become successful in math and college?

    _____

    _____

    _____

    _____

2. How can study breaks improve your learning?

    _____

    _____

    _____

    _____

3. What do you do when you don't want to study? How do you get started?

    _____

    _____

    _____

    _____

CHAPTER 5   REVIEW

1. Reading one page of a math text might take you _____ _____ _____.

2. After reading your textbooks, _____ important material can improve your ability to remember that information.

3. Write anticipated test questions in your _____ and indicate that they are your questions.

4. Checking your homework answers can improve your _____ and help you prepare for _____.

5. Matching:

(For students in a prealgebra class)

_____ Product   A. +

_____ Increased   B. −

_____ Sum   C. ×

_____ Per   D. ÷

_____ Is   E. =

_____ Reduced

_____ Remainder

(For students taking elementary algebra or a higher-level math course)

_____ Difference between x and 3   A. $10x$

_____ Quotient of x and 3   B. $x/3$

_____ Five times the sum of x and 2   C. $5(x - 4)$

_____ A number added to 5   D. $5(x + 2)$

_____ Five less than 10   E. $5 = x + 3$

_____ Ten times x F. $5 + x$

_____ Five is 3 more than a number   G. $10 - 5$

_____ Five times the difference of a number and 4   H. $x - 3$

6. Make sure you finish your homework by working problems you _____ do.

7. Using _____ _____ is a visual way to recall what you learned in math.

85

8. Using a _____ _____ is an auditory way to recall what you learned in math.

9. Most _____ occurs right after learning; a short review will help you _____ the new material.

10. A _____ _____ can improve your learning while helping you complete your homework assignments.

What is the most important information you learned from this chapter?

_____
_____
_____

_____

How can you immediately use it?

_____
_____
_____

_____

DAN'S TAKE REVIEW

1. What mistake did Dan make while scheduling his time for his first college math class?

_____
_____
_____

2. What was the problem with his logic that it would be better to cram four times during the semester rather than doing the homework on a daily basis?

_____
_____
_____

3. How could Dan avoid such problems in the future?

_____
_____
_____

How to Remember What You Have Learned

CHAPTER 6

To understand the learning process, you must understand how your memory works. You learn by conditioning and thinking. But memorization is different from learning. For memorization, your brain must perform several tasks, including receiving the information, storing the information, and recalling the information.

By understanding how your memory works, you will be better able to learn at which point your memory is failing you. Most students usually experience memory trouble between the time their brain receives the information and the time the information is stored.

There are many techniques for learning information that can help you receive and store information without losing it in the process. Some of these techniques may be more successful than others, depending on how you learn best.

87

DAN'S TAKE

Have you ever known someone who could learn something right away and then remember it perfectly forever? I had a friend like that in college. He could sit in on a lecture and remember the speech 2 years later without missing a beat.

Needless to say, most of us aren't like that. Most of our brains aren't strong enough to keep up with it all, so we hold onto only the information we care about and then filter the rest out of our minds. What results is a sizable working knowledge on things like album titles and baseball statistics, while more important information like the FOIL method might escape our memories far more easily. Personally, by my third year of college, I could spout off Ozzie Smith's career fielding percentage within a second of being asked, but if you were to ask me what a quotient was, I'd have to seriously think about it and might not be able to come up with an answer at all.

This process makes learning math, a sequential learning subject, more demanding than most other subjects. It's important when learning a step in a math problem to remember it for the rest of your college career.

So how do you remember math? I had this problem throughout high school and college. I eventually realized that the answer to this problem was simple: To remember math, you simply need to do it every day. I noticed that by breaking up my math homework and doing a little bit of it every day, I was able to keep the information fresh in my head. Usually this meant sitting down at my dorm room desk an hour or so before I would go to sleep for the night. This became a part of my nighttime routine and therefore relevant to my everyday life. The problems I was working with stayed in the front end of my working memory and were transferred to long-term memory rather than being buried in the back beneath random and unimportant factoids.

For the majority of us, remembering something takes effort. For me, remembering how to do math took a little bit of daily effort, and in the end it paid off.

6.1   How You Learn

Educators tell us that learning is the process of "achieving competency." More simply put, it is how you become good at something. The three ways of learning are by conditioning, by thinking, and by a combination of conditioning and thinking.

Learning by Conditioning and Thinking

Conditioning is learning things with a maximum of physical and emotional reaction and a minimum of thinking.

EXAMPLES

Repeating the word *pi* to yourself and practicing finding its symbol on a calculator are two forms of conditioned learning. You are learning using by your voice and your eye-hand coordination (physical activities), and you are doing very little thinking.

Thinking is defined as learning with a maximum of thought and a minimum of emotional and physical reaction.

EXAMPLE

Learning about *pi* by thinking is different from learning about it by conditioning. To learn *pi* by thinking, you would have to do the calculations necessary to arrive at the numeric value that the word *pi* represents. You are learning *by using your mind* (thought activities), and you are using very little emotional or physical energy to learn *pi* in this way.

The most successful way to learn is to combine thinking and conditioning. The best learning combination is to learn by thinking first and by conditioning second.

Learning by thinking means that you learn by

- observing
- processing
- understanding the information

How Your Memory Works

In order to remember information, we must receive and store it in a way that allows us to retrieve it. It is important to understand the process we must use to remember, because it can help us design more productive and time-efficient learning strategies. Briefly put, the memory process includes the following stages:

1. We receive information through our five senses: hearing, seeing, tasting, touching and smelling (**sensory input**).

2. Then, we briefly hold the image or sound of each sensory experience until it can be processed (**sensory register**).

3. After registering each sensory experience, we store the information that passes through the sensory register in **short-term memory**. We store small pieces of information in short-term memory for only a few seconds. Short-term memory is the ability to recall information immediately after it is given. If we do not do anything with it, such as put it in notes or learn it immediately, we lose this information. We can't put it into long-term memory.

4. Next, we use information we have stored in long-term memory along with new information that we have just processed to learn how to do math (**working memory**).

   Working memory is the ability to think about and use many pieces of information at the same time. It is like a mental chalkboard. It is also like the RAM in a computer, determining how many tasks can be done at the same time to produce an end product—understanding new math concepts. We also use reasoning skills to understand new concepts. **Reasoning** is thinking about these memories stored in long-term memory and comprehending what they mean. We also use reasoning and **long-term memory** to help us understand new concepts.

   The goal of all these steps is to get new information into long term memory. Long-term memory is a storehouse of information that is remembered for long periods of time.

5. The final goal for math students is to be able to show what they have learned—**memory output**. We recall what we have put into long-term memory and express this knowledge. It is the ability to go into the "storehouse of information," the long-term memory, and find what we need to complete a math problem.

Here is a story that describes the memory process.

On the second day of math class, Claudia, determined to start off strong, arrived at class early, enabling her to claim a desk where she could hear and make eye contact with the instructor. She wanted to make sure she caught everything he said. It worked. As the instructor began to discuss order of operations, she heard every word and saw everything he wrote on the board (sensory input and sensory register). Since she was able to pick up easily on what he was saying and writing, she was able to remember all the information until she could record it in her notes (short-term memory). Also going for her was the fact that she had attended a review workshop before the semester started, which helped her review everything from the course she had taken the previous semester. Her long-term memory was strong as a result. So, she was able to use her working memory and reasoning in class to understand the math, not just record the information in her notes. At the end of class, the instructor gave the students a practice quiz to see what they had learned during the past hour. She aced the quiz using her memory output.

What did Claudia do to make sure she worked through each stage of the memory process? You should take the same steps to improve your memory! Try the following suggestions provided in the table.

Memory Process	Strategies
Sensory input and sensory register	Sit where you can easily hear the professor and see what the professor writes on the board.
Short-term memory	Make sure you have a good system of taking notes. Get plenty of sleep so that you are alert.
Working memory	Make sure you complete all homework before going to class. Learn while you are doing your homework; do not just hurry through to get it done.
	Read the textbook before class to familiarize yourself with the vocabulary. You do not have to understand it all, but it will help your learning during class.
Long-term memory and reasoning	Whenever you are studying math, take time to memorize what you are learning and thinking. Use the study strategies mentioned later in this chapter to help you understand and memorize math concepts. Make mental cheat sheets to review constantly.
Memory output	Practice, practice, practice.
	Ask your instructor for practice tests. Rehearse the "Ten Steps to Better Test Taking" in Chapter 7. Get with a group of students, and make up practice test questions.

The consequences of something interfering with any one of these stages are substantial. Here are some possible consequences.

If you do not input and register the information:	There is nothing to work with. The rest of the class is confusing, as is your homework.
If you do not hold information in short-term memory long enough to record it in your notes:	You end up reteaching yourself when working on homework.
	Sometimes, you have to take more time and go to a tutor.
If you do not spend quality time doing homework, which limits your working memory that helps you understand the math:	You will not remember what you spent time doing.
	The information will not remain in long-term memory.
	You won't get a good grasp on what is going on.
If you do not spend time memorizing what you are learning:	You will forget and have to relearn and rememorize the night before the test.
	You won't do as well on the math tests.
	Learning new material will always be harder for you.
If you freeze up when trying to answer math problems (memory output):	You can't show what you really know.
	You get a lower score than you think you deserve.

It is important to take a few moments to talk more about memory output. Many students struggle with this. They know the information going into the test, but they forget it once the test is on their desks. Memory output can be blocked by three things:

1. **Insufficient processing of information into long-term memory.** For most students, this is a result of not studying correctly throughout the math unit. It is easier to fix than other memory blocks.

2. **Poor test-taking skills.** Most students have not mastered a system for taking tests. That is why an entire chapter is devoted to test taking. Once again, this is not difficult to fix.

3. **Test anxiety.** Test anxiety is a learned behavior that blocks memory. Students with test anxiety usually need to learn some strategies to help them manage their anxiety, as well as learn good test-taking strategies.

Many students who have trouble taking tests wait too long to get help. If you have a pattern of test anxiety in your past, get assistance before the first test. Success on the first test will help you overcome anxiety. Failure on the first test will only feed it. Be proactive and set up a plan for taking your first test.

What is great about our memory is that we can learn strategies to improve it. The next section will describe ways in which you can improve the memory process by practicing learning strategies based on your learning modalities and learning styles.

REMEMBER
Locating where your memory breaks down and compensating for those weaknesses will improve your math learning.

SECTION 6.1   REVIEW

1. Label the boxes in Figure 10 with the stages of memory.

FIGURE 10   Stages of Memory

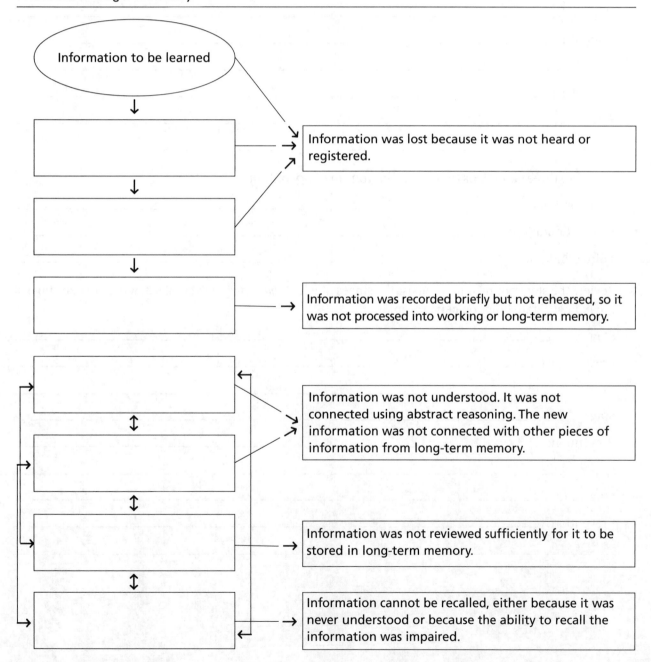

2. List, define, and give examples of each stage of memory.

    Stage 1: _____

    Example: _____

    Stage 2: _____

    Example: _____

    Stage 3: _____

    Example: _____

    Stage 4: _____

    Example: _____

    Stage 5: _____

    Example: _____

    Stage 6: _____

    Example: _____

    Stage 7: _____

    Example: _____

3. What are three conditions that can block your memory output?

    First Condition: _____

    Second Condition: _____

    Third Condition: _____

4. Review the stages of memory and list the stages at which *your* memory breaks down. For each stage of breakdown, state how you can prevent it.

    Stage: _____

    Prevention: _____

    _____

    Stage: _____

    Prevention: _____

    _____

    Stage: _____

    Prevention: _____

    _____

6.2   How to Use Learning Styles to Improve Memory

There are many different techniques that can help you store information in your long-term memory and reasoning. Using your learning sense or learning style and decreasing distractions while studying are very efficient ways to learn. Using your best *learning sense* (what educators call your "predominant learning modality") can improve how well you learn and enhance the transfer of knowledge into long-term memory/reasoning. Your learning senses are vision, hearing, touching, etc. Ask yourself if you learn best by watching (vision), listening (hearing), or touching (feeling).

REMEMBER

Learning styles are neither good nor bad and are based on genetics and environment. Knowing your best learning styles and using them effectively can dramatically improve your math learning and grades.

Visual (Watching) Learner

Knowing that you are a *visual math learner* can help you select the memory technique that will work best for you. Repeatedly reading and writing down math materials being studied is the best way for a visual learner to study.

Based on the Learning Styles Inventory, students who learn math best by seeing it written are *visual numerical learners*. If you are a visual numerical learner, you will learn best by following the 10 suggestions in Figure 11. Try as many of these suggestions as possible and then select and practice those that are most helpful.

A visual way to decrease distractions is by using the "my mind is full" concept. Imagine that your mind is completely filled with thoughts of learning math, and other distracting thoughts cannot enter. Your mind has one-way input and output, which responds only to thinking about math when you are doing homework or studying.

FIGURE 11   Visual Numerical Learners

1. Study a variety of written materials, such as additional handouts and math texts.
2. Play games with, and get involved in activities with, visible printed number problems.
3. Use visually orientated computer programs, CDROMs, and math websites.
4. Rework your notes using suggestions from this workbook.
5. Visualize numbers and formulas in detail.
6. Check out videocassette tapes from the math lab or learning resource center on campus.
7. Make 3" × 5" note (flash) cards, in color.
8. Use different colors of ink to emphasize different parts of each math formula.
9. Ask your tutor to *show* you how to do the problems instead of *telling* you how to do the problems.
10. Write down each problem step the tutor tells you to do. Highlight the important steps or concepts, that cause you difficulty.

Auditory (Hearing) Learner

If you are an *auditory learner* (one who learns best by hearing the information), then learning formulas is best accomplished by repeating them back to yourself or recording them on a tape recorder and listening to them. Reading out loud is one of the best auditory ways to get important information into long-term memory. Stating facts and ideas out loud improves your ability to think and remember. If you cannot recite out loud, recite the material to yourself, emphasizing the key words.

Based on the Learning Styles Inventory, students who learn math best by hearing it are *auditory numerical learners*. If you are an auditory numerical learner, you may learn best by following the 10 suggestions in Figure 12. Try as many of these suggestions as possible and then select and practice those that are most helpful.

An auditory way to improve your concentration is to become aware of your distractions and tell yourself to concentrate. If you are in a location

FIGURE 12   Auditory Numerical Learners

1. Say numbers to yourself or move your lips as you read problems.
2. Tape-record your class and play it back while reading your notes.
3. Read aloud any written explanations.
4. Explain to your tutor how to work math problems.
5. Make sure all important facts are spoken aloud with auditory repetition.
6. Remember important facts by auditory repetition.
7. Read math problems aloud and try solutions verbally and subverbally as you talk yourself through the problems.
8. Record directions to difficult math problems on audiotape and refer to them when solving those specific types of problems.
9. Have your tutor explain how to work problems instead of just showing you how to solve them.
10. Record math laws and rules in your own words, by chapters, and listen to them every other day (auditory highlighting).

where talking out loud will cause a disturbance, mouth the words "start concentrating" as you say them in your mind. Your concentration periods should increase.

Tactile/Concrete (Touching) Learner

A *tactile/concrete learner* needs to feel and touch the material to learn it. Tactile/concrete learners, who are also called *kinesthetic learners*, tend to learn best when they can concretely manipulate the information to be learned. Unfortunately, most math instructors do not use this learning sense. As a result, students who depend heavily on feeling and touching for learning will usually have the most difficulty developing effective math learning techniques. This learning style creates a problem with math learning because math is more abstract than concrete. Also, most math instructors are visual abstract learners and have difficulty teaching math tactilely. Ask for the math instructors and tutors who give the most practical examples and who may even "act out" the math problems.

As mentioned before, a tactile/concrete learner will probably learn most efficiently by hands-on

FIGURE 13   The FOIL Method

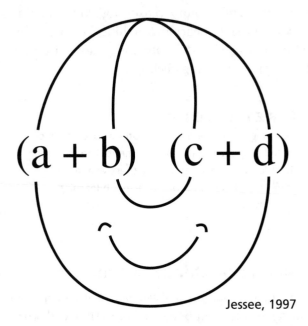

Jessee, 1997

The FOIL Method is used to remember the procedure for multiplying two binomials. The letters in FOIL stand for First, Outside, Inside, and Last. To use the FOIL Method, multiply

- the First terms ((a) (c)),
- the Outside terms ((a) (d)),
- the Inside terms ((b) (c)),
- the Last terms ((b) (d)),
- and then combine the four products.

To learn the FOIL Method, trace your finger along the FOIL route.

learning. For example, if you want to learn the FOIL method, you would take your fingers and trace the "face" to remember the steps. See Figure 13 (The FOIL Method). Also, learning is most effective when physical involvement with manipulation is combined with sight and sound. For example, as you trace the face you also say the words out loud.

Based on the Learning Styles Inventory, tactile/concrete learners best learn math by manipulating the information that is to be taught. If you are a tactile/concrete learner, you may learn best by following the 10 suggestions in Figure 14. Try as many of these suggestions as possible and then select and practice the ones that help the most. If you do not have these manipulatives or don't know how to use them, ask the math lab supervisor or instructor if they have any manipulative materials or models. If the math lab does not have any manipulative materials, you may have to ask for help to develop your own.

Tactile/concrete learners can also use graphing calculators to improve their learning. Entering the keystrokes makes it easier to remember how to solve the problems. This practice is also an excellent way to remember how to solve the problem when using a calculator while taking a test.

FIGURE 14 Tactile/Concrete Learners

1. Cut up a paper plate to represent fractions of a whole.
2. Fold a piece of paper several times and cut along the fold marks to represent fractions of a whole.
3. In order to understand math concepts, ask to be shown how to use algebra tiles as manipulatives.
4. Try to use your hands and body to "act out" a solution. For example, you may "become" the car in a rate-and-distance word problem.
5. Obtain diagrams, objects, or manipulatives and incorporate activities such as drawing and writing into your study time. You may also enhance your learning by doing some type of physical activity such as walking.
6. Try to get involved with at least one other student, a tutor, or an instructor who uses manipulatives to help you learn math.
7. Ask to use the Hands-on Equations Learning System, which uses manipulatives, to learn basic algebra. You can go to their website (www.Borenson.com) to learn more about this system and other systems that can help you learn math.
8. Go to one of the "learning stores" in shopping malls to see if they have manipulatives.
9. Go to a K–12 learning resource store to see if they have manipulatives, such as magnetic boards with letters and numbers.
10. Talk to the coordinator of students with disabilities to see if they use manipulatives when tutoring their students.

Another way tactile/concrete learners can learn is to trace graphs with their fingers when they appear on the calculator. They should read out loud and trace every equation to "feel" how the graph changes for different equations. For example, if you add 2 to one side of an equation, move your finger to where the graph changes and say out loud how much it moves.

A tactile/concrete way to improve your study concentration is by counting the number of distractions during each study session. Place a sheet of paper by your book when doing homework. When you catch yourself not concentrating, write the letter C on the sheet of paper. This will remind you to concentrate and get back to work. After each study period, count up the number of Cs and watch the number decrease over time.

Social Individual Learner

If you are a *social individual learner,* you may learn math best by learning it individually. You may learn best by yourself, working with computer programs and an individual tutor. In some cases, social individual learners may have to meet in groups to develop practice tests but leave socializing to a minimum. If you are a social individual learner and a visual learner, the computer may be one of the best learning tools available. If you are a social individual learner, based on the Learning Styles Inventory, you may learn best by following the eight suggestions in Figure 15. Try as many of these suggestions as possible and then select those that are most helpful.

A problem that social individual learners may encounter is working too long on problems for which they could have received help. Social individual learners must understand that getting help is okay, especially if it saves study time and makes their studying more efficient.

Social Group Learners

If you are a *social group learner* (one who best learns in a group), then learning math may best be done in study groups and in math classes that offer collaborative learning (group learning). Social group learners may learn best by discussing information.

They can usually develop their own study groups and discuss how to solve problems over the phone.

FIGURE 15 Social Individual Learners

1. Study math, English, and your other subjects alone.
2. Utilize videocassette tapes or auditory tapes to learn by yourself.
3. Prepare individual questions for your tutor or instructor.
4. Obtain individual help from the math lab or hire your own tutor.
5. Set up a study schedule and a study area so that other people will not bother you.
6. Study in a library or in some other private, quiet place.
7. Use group study times only as a way to ask questions, obtain information, and take pretests on your subject material.
8. Use math websites to help you learn.

FIGURE 16 Social Group Learners

1. Study math, English, and your other subjects in a study group.
2. Sign up for math course sections that offer cooperative learning (learning in small groups).
3. Review your notes with a small group.
4. Obtain help in the math lab or other labs where you can work in group situations.
5. Watch math videocassette tapes with a group and discuss the subject matter.
6. Listen to audiocassette tapes on the lecture and discuss them with the group.
7. Obtain several "study buddies" so you can discuss with them the steps to solving math problems.
8. Form a study group. Each member should bring ten test questions with explanations on the back. The group should complete all the test questions.

If you are a social group learner and an auditory learner, then you definitely learn best by talking to people. If you are a social group learner, based on the Learning Styles Inventory, you may learn best by following the eight suggestions in Figure 16. Try as many of these suggestions as possible and then select and practice those that are most helpful.

A learning problem that a social group learner may have is talking too much about other subjects when in a study group. This is called being off task. You may want to have a group member serve as a discussion monitor and let the other students know when they need to get back on task. Also, social group learners need to know that they still must study math individually to be successful. During these individual study sessions, they should prepare questions for the group.

Multiple Senses

If you have difficulty learning material through one sense (learning style), you might want to try

learning material through two or three senses. Involving two or more senses in learning improves your learning and remembering. Review the figures in this section on learning styles and, whenever possible, combine two or more learning styles to improve learning.

If your primary sense is visual and your secondary sense is auditory, you may want to write down equations while saying them out loud to yourself. Writing and reciting the material at the same time combines visual, auditory, and (to some extent) tactile/concrete styles of learning.

Studying with a pen or highlighter is a visual as well as a tactile/concrete way to improve your concentration. Placing the pen or highlighter in your hand and using it will force you to concentrate more on what you are reading. After you write and recite the material back to yourself, do it 5 or 10 more times to over learn it.

SECTION 6.2 REVIEW

1. List five ways visual learners can improve their memory.

First Way: _____

Second Way: _____

Third Way: _____

Fourth Way: _____

Fifth Way: _____

2. List five ways auditory learners can improve their memory.

First Way: _____

Second Way: _____

Third Way: _____

Fourth Way: _____

Fifth Way: _____

3. List five ways tactile/concrete learners can improve their memory.

First Way: _____

Second Way: _____

Third Way: _____

Fourth Way: _____

Fifth Way: _____

4. List three ways social individual learners can improve their memory.

First Way: _____

Second Way: _____

Third Way: _____

5. List three ways social group learners can improve their memory.

First Way: _____

Second Way: _____

Third Way: _____

6. What is your best modality learning style for math? Visual, auditory, or tactile/concrete?

_____

_____

7. What is your second-best modality learning style for math?

_____

8. Are you more of an individual or a group learner? Why?

_____

_____

_____

9. How can you use multiple senses to improve your memory?

_____

_____

_____

10. What is your best combination of learning styles to learn math? _____

_____

_____

Give an example of how you can combine your learning styles to learn a math concept. (If you cannot think of a concept, ask your instructor for one.)

_____

_____

_____

6.3 How to Use Memory Techniques

There are many different techniques that can help you store information in your long-term memory: a positive attitude about studying, selective learning, organization, visual imagery, mnemonic devices, and acronyms.

A Good Study/Math Attitude

A positive attitude about studying will help you concentrate and improve your retention of information. This means that you should have at least a neutral math attitude (you neither like nor dislike it) and should reserve the right to learn to like math. If you still don't like math, just pretend that you do while studying it. View studying as an opportunity to learn rather than an unpleasant task. Tell yourself that you can learn the material, and it will help you pass the course and graduate.

Be a Selective Learner

Being selective in your math learning will improve your memory. Prioritize the materials you are studying; decide which facts you need to know and which ones you can ignore. Narrow down information into laws and principles that can be generalized. Learn the laws and principles 100%.

EXAMPLE

If you have been given a list of math principles and laws to learn for a test, put each one on an index card. As you go through them, create two piles: an "I already know this" pile and an "I don't know this" pile. Then study *only* the cards in the "I don't know this" pile. Study the principles on these cards until you have memorized them and understand them completely.

Become an Organizer

Organizing your math material into idea/fact clusters or groups will help you learn and memorize it. Grouping similar material in a problem log or calculator log is an example of categorizing information. Do not learn isolated facts; always try to connect them with other similar material.

Use Visual Imagery

Using mental pictures or diagrams to help you learn the material is especially helpful for visual learners and for students who are right-hemisphere dominant (who tend to learn best by visual and spatial methods). Mental pictures and actual diagrams involve 100% of your brain power. Picturing solution steps can help you solve difficult math problems in your mind.

> ### EXAMPLE
>
> Use the FOIL method (Figure 13) to learn visually how to multiply binomials. Memorize the face until you can sketch it from memory. If you need to use it during a test, you can sketch the face on your scratch paper and refer to it.

Make Associations

Association learning can help you remember better. Find a link between new facts and some well-established old facts and study them together. The recalling of old facts will help you remember the new facts and strengthen a mental connection between the two. Make up your own associations to remember math properties and laws.

> ### REMEMBER
>
> The more ridiculous the association, the more likely you are to remember it.

> ### EXAMPLES
>
> When learning the *commutative property*, remember that the word *commutative* sounds like the word *community*. A community is made up of different types of people who could be divided into an *a* group and a *b* group. However, in a community of *a* people and *b* people, it does not matter if we count the *a* people first or the *b* people first; we still have the same total number of people in the community. Thus, $a + b = b + a$.
>
> When learning the *distributive law of multiplication over addition*, such as $a(b + c)$, remember that *distributive* sounds like *distributor*, which is associated with giving out a product. The distributor *a* is giving its products to *b* and *c*.

Use Mnemonic Devices

The use of mnemonic devices is another way to help you remember. Mnemonic devices are easily remembered words, phrases, or rhymes associated with difficult-to-remember principles or facts.

> ### EXAMPLE
>
> Many students become confused by the order of operations. These students mix up the order of the steps in solving a problem, such as dividing before adding the numbers in parentheses. A mnemonic device that can help you remember the order of operations is "Please Excuse My Dear Aunt Sally." The first letter in each of these words represents the math function to be completed from the first to the last. Thus, the order of operations is Parentheses (**P***lease*), Exponents (**E***xcuse*), Multiplication (**M***y*), Division (**D***ear*), Addition (**A***unt*), and Subtraction (**S***ally*). Remember to multiply and/or divide whatever comes first, from left to right. Also, add or subtract whatever comes first, from left to right.

Use Acronyms

Acronyms are another memory device that can help you learn math. Acronyms are word forms created from the first letters of a series of words.

> ### EXAMPLE
>
> FOIL is one of the most common math acronyms. FOIL is used to remember the procedure for multiplying two binomials. Each letter in the word *FOIL* represents a math operation. **FOIL** stands for **F**irst, **O**utside, **I**nside, and **L**ast, as it applies to multiplication of two binomials such as $(2x + 3)(x + 7)$. The **F**irst quantities are $2x$ (in the first expression) and x (in the second expression). The **O**utside quantities are $2x$ (in the first expression) and 7 (in the second expression). The **I**nside quantities are 3 (in the first expression) and x (in the second expression). The **L**ast quantities are 3 (in the first expression) and 7 (in the second expression). This results in
>
> **F** $(2x)(x)$ + **O** $(2x)(7)$ + **I** $(3)(x)$ + **L** $(3)(7)$.
>
> Do the multiplication to get $2x^2 + 14x + 3x + 21$, which adds up to $2x^2 + 17x + 21$. See Figure 13 (The FOIL Method).

SECTION 6.3   REVIEW

1. How can a good math attitude help you learn?

   _____

   _____

2. Give an example of being a selective learner in your math class.

   _____

   _____

3. From your current math lessons, make up and explain one association remembering device that is not in this workbook.

   _____

   _____

4. For your next major math test, make up and explain one mnemonic device that is not in this workbook.

   _____

   _____

5. For your next major math test, make up and explain one acronym that is not in this workbook.

   _____

   _____

6.4   How to Develop Practice Tests

Developing a practice test is one of the best ways to evaluate your memory and math skills before taking a real test. You want to find out what you do not know *before* the real test instead of *during* the test. Practice tests should be as real as possible and should include time constraints.

You can create a practice test by reworking all the problems that you have recorded in your problem log since your last test. Another practice test can be developed using every other problem in the textbook chapter tests. Further, you can use the solutions manual to generate other problems with which to test yourself. You can also use old exams from the previous semester. Check to see if your math lab/LRC or library has tests on file from previous semesters, or ask your instructor for other tests.

For some students, a better way to prepare for a test is the group method.

EXAMPLE

Hold a group study session several days before the test. Have each student prepare a test with 10 questions. On the back of the test, have the answers listed, worked out step by step. Have each member of the study group exchange his or her test with another member of the group. Once all the tests have been completed, have the author of each test discuss with the group the procedures used to solve the problems in his or her test.

If group work improves your learning, you may want to hold a group study session at least once a week. Make sure the individual or group test is completed at least 3 days before the real test.

Completing practice math tests will help you increase your testing skills. It will also reveal your test problem weaknesses in enough time for you to learn how to solve the problems before the real test. If you have difficulty with any of the problems during class or after taking the practice test, be sure to see your tutor or instructor.

After taking the practice test(s), you should know what types of problems you do not understand (and need to study) and what is likely to be on the actual test. Put this valuable information on one sheet of paper. This information needs to be

understood and memorized. It may include formulas, rules, or steps in solving problems.

Use the learning strategies discussed in this chapter to remember this information. A good example of how this information should look is what students might call a mental "cheat sheet." Obviously, you cannot use the written form of this sheet during the real test.

If you cannot take a practice test, put down on your mental cheat sheet the valuable information you will need for the test. Work to understand and memorize your mental cheat sheet. Chapter 7, "How to Improve Your Math Test-Taking Skills," will discuss how to use the information on the mental cheat sheet—*without cheating.*

SECTION 6.4 REVIEW

1. List three different ways you can make up a practice test.

   First Way: _____

   Second Way: _____

   Third Way: _____

2. List 10 different problems you would put on your practice test for the next exam.

   First Problem: _____

   Second Problem: _____

   Third Problem: _____

   Fourth Problem: _____

   Fifth Problem: _____

   Sixth Problem: _____

   Seventh Problem: _____

   Eighth Problem: _____

   Ninth Problem: _____

   Tenth Problem: _____

3. Give these 10 problems to your study buddy and get 10 problems from him or her. Compare your answers with those of your study buddy. How many did you get right? _____ How can you correct the wrong answers?

   _____

   _____

   _____

   _____

6.5   How to Use Number Sense

When taking your practice tests, you should use "number sense," or estimations, to make sure your answers are reasonable. Number sense is common sense applied to math. Number sense is the ability to see if your answer makes sense without using algorithms. (Algorithms are the sequential math steps used to solve problems.) The following examples demonstrate solving a math problem (from a national math test given to high school students) using algorithms and number sense.

EXAMPLE

Solve 3.04 × 5.3. Students used algorithms to solve this problem by multiplying the number 3.04 by 5.3, in sequence. Seventy-two percent of the students answered the problem correctly using algorithms.

EXAMPLE

Estimate the product 3.04 × 5.3, and choose from the following answers.

    (A) 1.6   (C) 160

    (B) 16   (D) 1600

Only 15% of the students chose B, which is the correct answer. Twenty-eight percent of the students chose A. Using *estimation* to solve the answer, a whopping 85% of the students got the problem wrong.

These students were incorrectly using their "mental chalkboard" instead of using number sense. In using number sense, you would multiply the numbers to the left of the decimal point in each number to estimate the answer. To estimate the answer, then, you would multiply 3 (the number to the left of the decimal in 3.04) by 5 (the number to the left of the decimal in 5.3) and expect the answer to be a little larger than 15.

It appears that the students' procedural processing (the use of algorithms) was good, but when asked to solve a nonroutine problem using estimation (which is easier than using algorithms), the results were disappointing.

Another example of using number sense, or estimation, is rounding.

Taking the time to estimate the answer to a math problem is a good way to check your answer. Another way to check your answer is to see if it is reasonable. Many students forget this important step and get the answer wrong. This is especially true of word or story problems.

EXAMPLE

Solve 48 + 48 by rounding. Rounding means mentally changing the number (up or down) to make it more manageable to you, without using algorithms. By rounding, 48 becomes 50 (which is easier to work with). 50 + 50 = 100. If the choices for answers were 104, 100, 98, and 96, you would then subtract 4 from 100 (since each number was rounded up by 2) and you would get 96.

Also, when dealing with an equation, make sure to substitute the answer back into the equation to see if one side of the equation equals the other. If the two sides are not equal, you have the wrong answer. If you have extra time after you have completed a test, you should check answers using this method.

EXAMPLES

When solving a rate-and-distance problem, use your common sense to realize that a car cannot go 500 miles per hour to catch another car. However, the car can go 50 miles per hour.

The same common sense rule applies to age problems, where the age of a person cannot be 150 years, but it can be 15 years.

Furthermore, in solving equations that are not word problems, *x* is *usually* a number that is less than 20. When you solve a problem for *x* and get 50, then this isn't reasonable, and you should recheck your calculations.

EXAMPLE

In solving the equation $x + 3 = 9$, you calculated that $x = 5$. To check your answer, substitute 5 for x and see if the problem works out correctly. $5 + 3$ does not equal 9, so you know you have made a mistake and need to recalculate the problem. The correct answer, of course, is $x = 6$.

 REMEMBER

Number sense is a way to get more math problems correct by estimating your answer to determine if it is reasonable.

SECTION 6.5   REVIEW

1. Give an explanation of number sense.

   _____

   _____

2. How does number sense compare with common sense?

   _____

   _____

3. Give an example of a recent math problem you failed to solve correctly as a result of number sense.

   _____

   _____

4. List two ways you can improve your number sense.

   First Way: _____

   Second Way: _____

1. The way you receive information is through your five senses, which are _____, _____, _____, _____, and _____.

2. The main problem most students have is converting learned material into _____ _____ and _____ it.

3. Repeatedly reading and writing math material is one of the best ways for a _____ learner to study.

4. Reading _____ _____ is one of the best auditory ways to learn material.

5. A _____/ _____ learner needs to feel and touch material to learn it.

6. Being a selective learner means _____ the material to study and learning the laws and principles _____ percent.

7. Mnemonic devices are easy-to-remember _____, _____, or _____ associated with difficult-to-remember principles or facts.

8. _____ are word forms created from the first letters of a series of words.

9. The reason to develop a practice test is to find out what you do not know _____ the test instead of _____ the test.

10. Social group learners benefit from forming _____ _____.

What is the most important information you learned from this chapter?

_____

_____

_____

How can you immediately use it?

_____

_____

_____

DAN'S TAKE REVIEW

1. Why is it that Dan could remember detailed baseball statistics but could not remember something as simple as the FOIL method?

_____

_____

_____

2. How did Dan solve this problem and eventually learn how to remember math?

_____

_____

_____

3. If the answer to remembering math properly is daily use, when would be the best time in your daily schedule to sit down and study?

_____

_____

_____

LECTURE NOTES

How to Improve Your Math Test-Taking Skills

CHAPTER 7

Taking a math test is different from taking tests in other subjects. Math tests require you not only to recall the information, but also to apply the information. Multiple-choice tests, for example, usually test you on recall, and if you do not know the answers, you can guess.

Math tests build on each other, whereas history tests often do not test you on previous material. Most math tests are conceptional tests, where you cannot memorize the answers; most social science tests are designed to test long-term memory and for everyone to finish.

Math test preparation and test-taking skills are different from preparation and skills needed for other tests. You need to have a test-taking plan and a test analysis plan to demonstrate your total knowledge on math tests. Students with these plans make better grades compared with students without them. Math instructors want to measure your math knowledge, not your test-taking skills.

107

The moment the professor hands out a math test, everyone dives in, frantically jotting down answers as fast as they can. The first few questions are easy, giving the test takers a false sense of security heading into the second page. At this point, those who know what they are doing begin to surge ahead, panicking their neighbors and pressuring them to hurry through their tests to keep up.

Then the first person finishes, and one by one, those taking their time watch their classmates leave. The few students that remain are left with a sense of failure before they even turn in their tests.

I don't understand why college students believe that a test is a race. Everyone knows that those who finish first don't always score the highest. But for some reason, we still feel the need to avoid the dreaded "last person to finish" label.

Math and speed do not mix well together. In fact, trying to finish a test quickly often causes mistakes. Speed "kills" math grades. Even if you know the steps, a small error can make you get a question wrong. It might be

that you just carried a number incorrectly. These mistakes happen when you rush through a test.

In high school I often fell into this trap. I would study and prepare for a test and then be shocked when I got a C. During the test, as I watched my peers pick up their things and leave the class, I would panic. I, like so many others that age, was terrified that I might appear stupid if I were the last person to leave the room.

Everything changed when I realized that I didn't need to be the first person out the door after a test. By taking my time, I was able to catch all the little mistakes that were making me lose points. I could leave with the confidence that I honestly represented my knowledge of the material at hand.

It is very important that you not feel pressured during a math test. When people finish before you, ignore them. What's the big deal? It's not fair to measure yourself by how quickly those around you work. Instead of worrying about getting things done quickly, worry about getting them done correctly and making a good grade and thus passing the course.

7.1 Why Attending Class and Doing Your Homework May Not Be Enough to Pass

Most students and some instructors believe that attending class and doing all the homework ensures an A or a B on tests. This is far from true. Doing all the homework and getting the correct answers is very different in many ways from taking tests:

1. While doing homework, there is little anxiety. A test situation is just the opposite.

2. You are not under a time constraint while doing your homework; you may have to complete a test in 55 minutes or less.

3. If you get stuck on a homework problem, your textbook, notes, and Websites are there to assist you. This is not true for most tests.

4. Once you learn how to do several problems in a homework assignment, the rest are similar. In a

test, the problems may be in random order and measure different concepts.

5. In doing homework, you have the answers to at least half the problems in the back of the text and answers to all the problems in the solutions guide. This is not true for tests.

6. While doing homework, you have time to figure out how to use your calculator. During the test, you can waste valuable time figuring out how to use your calculator.

7. When doing homework, you can call your study buddy or ask the tutor for help, something you cannot do during the test.

8. While doing your homework, you can go to Websites for online tutoring to help you solve problems you do not understand.

9. While doing online homework, you have the hints if you get stuck and other resources. These resources are not available during tests.

10. While doing your homework in the math lab or Learning Resource Center, when students leave it does not bother you. In a test situation when students get up and leave, it may cause anxiety and worry and make you want to leave before checking your problems.

Do not develop a false sense of security by believing you can make an A or a B by just doing your homework. Tests measure more than just your math knowledge.

The General Pretest Rules

General rules are important when taking any type of test:

1. *Get a good night's sleep before taking a test.* This is true for the ACT, the SAT, and your math tests. If you imagine you are going to cram all night and perform well on your test with 3 to 4 hours of sleep, you are wrong. It would be better to get 7 or 8 hours of sleep and be fresh enough to use your memory to recall information needed to answer the questions.

2. *Start studying for the test at least 3 days ahead of time.* Make sure you take a practice test to find out, before the test, what you do not know. Review your problem log and work the problems. Review the concept errors you made on the last test. (How to identify and correct your concept errors will be discussed later in this chapter.) Meet with your instructor or tutor for help on those questions you cannot solve.

3. *Review only the material you have already learned the night before a test.*

4. *Make sure you know all the information on your mental cheat sheet.* Review your notebook and glossary to make sure you understand the concepts. Work a few problems and recall the information on your mental cheat sheet right before you go to bed. Go directly to bed; do not watch television, listen to the radio, or party. While you are asleep, your mind will work on and remember the last thing you did before you went to sleep.

5. *Get up in the morning at your usual time and review your notes and problem log.* Do not do any new problems, but make sure your calculator is working.

SECTION 7.1 REVIEW

1. List three reasons why only attending class and doing your homework may not be enough to pass your math course.

First Reason: _____

_____

_____

_____

Second Reason: _____

_____

_____

_____

Third Reason: _____

_____

_____

_____

_____

2. List and explain three general pretest rules that best apply to you.

First Rule: _____

_____

_____

_____

_____

Second Rule: _____

_____

_____

_____

Third Rule: _____

_____

_____

_____

_____

- -

PRETEST CHECKOFF LIST

Just like a pilot who conducts a preflight check list, you can do a pretest check list in order take off safely and not "crash" on your test. Complete as many of the items on the checkoff list as you can 2 days or more *before* the test.

Reducing Test Anxiety

Students who have test anxiety need to practice the relaxation techniques that they will use before and during the test. Reducing test anxiety can help students recall information for the test.

I have practiced my best short-term relaxation technique.   Yes _____   No _____   *Don't have test anxiety*
_____

Students with test anxiety usually say very negative statements to themselves, which intensifies their test anxiety further.

What I usually say to myself during a test is not positive.   Yes ____   No ____
What are these negative statements?

1. _____

2. _____

3. _____

I have written and practiced saying positive   Yes _____ No _____
statements to replace the negative ones.
What are these positive statements?

1. _____

2. _____

3. _____

Reviewing Notes and Problem Log

Understanding math vocabulary is key to learning math. Talking about the meaning and uses for vocabulary words helps to review the key concepts you must remember for a test. To review math vocabulary, use your math glossary and/or the key words in your note-taking system.

I have reviewed my math glossary or key words at least   Yes _____ No _____
three or four times a week.   *Did not develop a glossary or set of key words _____*

I reviewed my math glossary or key words at least once   Yes _____ No _____
a day starting 2 days before the test.   *Did not develop a glossary or set of key words _____*

Reviewing all your notes before the test is a good strategy to help you remember math concepts and how to do math problems.

I have reviewed all my notes 2 days before the test.   Yes _____ No _____
                                                               Did not take notes _____

Reviewing your problem log of instructor examples is another way to understand math concepts and problem-solving strategies.

I have reviewed my problem log examples by 2 days before the test.   Yes _____ No _____
                                                               Did not develop a problem log _____

Reviewing the Textbook

Reviewing the examples in the textbook is another way to refresh your math skills.

I have reviewed the examples in the textbook 2 days before the test.   Yes _____ No _____

One of the strategies of reading a textbook is to develop anticipated test questions.

I reviewed my anticipated test questions.   Yes _____ No _____
                                                               Did not develop anticipated test questions _____

Reviewing Homework

Reviewing your homework problems is a good way to help remember how to do the problems.

I reviewed homework problems 2 days before the test.   Yes _____ No _____
                                                               Did not do the homework problems _____

To study productively, it is important to identify the problems and concepts you do not know well. Instead of repeatedly reviewing what you already know well, focus your energy on improving your comprehension and ability to complete the more challenging problems.

I have identified what I do not know well enough for the test   Yes _____ No _____
at least a week before the test.

Memory Techniques

Memory techniques can be used for hard-to-remember rules, concepts, or problems. Memory techniques, which include repeating concepts or problems out loud, making associations, using mnemonic devices, and using acronyms, will improve your test-taking skills.

I have developed memory techniques for my test.   Yes _____ No _____

Reviewing Concept Errors

One of the six types of test-taking errors, which are discussed later in this chapter, is concept error. Concept errors from previous tests need to be reviewed in order not to make the same mistakes again. Reviewing your concept errors can improve your understanding of mathematics.

Did you review your concept errors from previous tests?   Yes _____ No _____
                                                                     Did not record concept errors _____

Using Practice Tests

Practice tests help you find out what you don't know before the test instead of during the test. The best way to develop a practice test is with other students, because you can exchange questions. Developing practice tests can improve your ability to solve math problems on the real test.

I worked with other students to develop and take a practice test   Yes _____ No _____
at lease 2 days before the test.   *Did not develop practice tests* _____

Taking instructor-made practice tests can also help you to find out what problems you need to work on before, instead of during, the test. Using practice tests can help you find out what you need to study before the test.

I have worked the problems on the instructor's practice test.   Yes _____ No _____
                                                                          There was no instructor practice test _____

Memory Data Dump

The information from a memory data dump can be used to help you during the test. Using a memory data dump on your test can improve your test scores. You can also make up memory techniques based on the problems that you missed on the practice tests.

I have developed memory devices to use during a data dump.   Yes _____ No _____

Test-Taking Strategies

Some students use the "Ten Steps to Better Test-Taking" as a test-taking guide while others develop their own test-taking strategies. Either technique can improve your test-taking strategy.

I have developed and practiced my test-taking strategy.   Yes _____ No _____

Are you wondering why we are asking you to complete the preceding items 2 days or more before the test instead of the night before the test? If you get stuck on a problem, find you don't understand a concept, or cannot develop one of the strategies, then you will still have at least 1 day to get help before the test. Remember, the night before the test you need to be reviewing material you already have learned—not learning new material. The most productive way to complete this checklist is to include these steps in your study time every day.

7.2   Ten Steps to Better Test Taking

You need to have a game plan to take a math test. This plan is different from plans for taking history, English, humanities, and some science tests. The game plan is to get the most points in the least amount of time. Many students lose test points because they use the wrong test-taking strategies for math. By following these ten steps you can demonstrate more knowledge on the test and get more problems right.

Step 1   *Use a memory data dump.* When you get your test, turn it over and write down the information that you might forget. Remember, this is your mental cheat sheet that you should have already made while preparing for the test. Your mental cheat sheet has now turned into a mental list, and writing down this information is not cheating. Do not put your name on the test, do not skim it, just turn it over and write down those facts, figures, and formulas from your mental cheat sheet or other information you might not remember during the test. This is called your *first memory data dump.* The data dump provides memory cues for test questions.

> **EXAMPLE**
>
> It might take you a while to remember how to do a coin problem. However, if you had immediately turned your test over and written down different ways of solving coin problems, it would be easier to solve the coin problem.

Step 2   *Preview the test.* Put your name on the test and start previewing. Previewing the test requires you to look through the entire test to find different types of problems and their point values. Put a mark by each question that you can do without thinking. These are the questions that you will solve first.

Step 3   *Do a second memory data dump.* The second data dump is for writing down material that was jarred from your memory while previewing the test. Write this information on the back of the test.

Step 4   *Develop a test progress schedule.* When you begin setting up a test schedule, determine the point value for each question. Some test questions might be worth more points than others.

In some tests, word problems are worth 5 points and other questions might be worth 2 or 3 points. You must decide the best way to get the most points in the least amount of time. This might mean working the questions worth 2 or 3 points first and leaving the more difficult word problems for last.

Decide how many problems should be completed halfway through the test. You should have more than half the problems completed by that time.

Step 5   *Answer the easiest problems first.* Solve, in order, the problems you marked while previewing the test. Then review the answers to see if they make sense. Start working through the test as fast as you can while being accurate. Answers should be reasonable.

> **EXAMPLE**
>
> The answer to a problem asking you to find the area of a rectangle cannot be negative, and the answer to a land-rate-distance problem cannot be 1000 miles per hour.

Clearly write down each step to get partial credit. Even if you wind up with an incorrect answer, you might get some credit. In most math tests, the easier problems are near the beginning of the first page; you need to answer them efficiently and quickly. This will give you both more time for the harder problems and time to review.

Step 6   *Skip difficult problems.* If you find a problem that you do not know how to work, read it twice and automatically skip it. Reading it twice will help you understand the problem and put it into your working memory. While you are solving other problems, your mind will still be working on that problem. A difficult problem could be a type of problem you have never seen before or a problem in which you get stuck on the second or third step. In either case, skip the problem and go on to the next one.

Step 7   *Review the skipped problems.* When working the skipped problems, think how you have solved other similar problems. Also try to remember how the instructor solved that type of problem on the board.

While reviewing skipped problems, or at any other time, you may have the "Aha!" response. The "Aha!" response is your remembering how to do a skipped problem. Do not wait to finish your current problem. Go to the problem on which you had the "Aha!" and finish that problem. If you wait to finish your current problem, your "Aha!" response could turn into an "Oh, no!" response.

Step 8 *Guess at the remaining problems.* Do as much work as you can on each problem, even if it is just writing down the first step. If you cannot write down the first step, rewrite the problem. Sometimes rewriting the problem can jar your memory enough to do the first step or the entire problem. This is particularly true for tactile/concrete learners, for whom writing can trigger the memory process of solving the problem. Remember, you did not learn how to solve math problems with your hands tied behind your back. If you leave the problem blank, you will get a zero. Do not waste too much time on guessing or trying to work the problems you cannot do.

Step 9 *Review the test.* Look for careless errors or other errors you may have made. Students usually lose 2 to 5 test points on errors that could have been caught in review. Do not talk yourself out of an answer just because it may not look right. This often happens when an answer does not come out even. It is possible for the answer to be a fraction or a decimal.

REMEMBER

Research reveals that the odds of changing a right answer to a wrong answer are greater than the odds of changing a wrong answer to a right one.

Step 10 *Use all the allowed test time.* Review each problem by substituting the answer back into the equation or doing the opposite function required to answer the question. If you cannot check the problem in either of these two ways, rework the problem on a separate sheet of paper and compare the answers. Do not leave the test room until you have reviewed each problem two times or until the bell rings.

Even though we encourage students to work until the end of the test period, most students leave the classroom before the end of the period. These students state that even though they know they should use all the test time, they cannot stay in the room until the end of the test time. These students also know that their grades would probably improve if they kept checking their answers or kept working the problems they were having difficulty with. After talking to hundreds of these students, I discovered two different themes for leaving the classroom early. First, test anxiety gets so overwhelming that they cannot stay in the room. The relief from the test anxiety (leaving the room) is worth more than getting a better grade. If you are one of these students, you must learn how to decrease your test anxiety by following the suggestions in Chapter 3 or by using the *How to Reduce Test Anxiety* CD. Don't let test anxiety control your grades!

The other reason for leaving the test early is that they do not want to be the last student or one of the last few students to turn in their tests. They still believe that students who turn their tests in last are "dumb and stupid." These students also believe that students who turn their tests in first make As and Bs and students who turn their tests in last make Ds and Fs. If you are one of these students, you don't need to care about what other students think about you (it's usually wrong anyway). YOU need to fight the urge to leave early and use all the test time. Remember, passing mathematics is the best way to get a high-paying job and support yourself or your family.

REMEMBER

There is no prize for handing your test in first, and some students who turn their papers in last make As.

Handing in your scratch paper with your math test has several advantages:

- If you miscopied an answer from the scratch paper, you will probably get credit for the problem.

- If you get the answer incorrect due to a careless error, your work on the scratch paper could give you a few points.

- If you do get the problem wrong, it will be easier to locate the errors when the instructor reviews the test. This will prevent you from making the same mistakes on the next math test.

REMEMBER

Handing in your scratch paper may get you extra points or improve your next test score.

SECTION 7.2   REVIEW

1. List and explain the ten steps to better test taking.

Step 1: _____

_____

Step 2: _____

_____

Step 3: _____

_____

Step 4: _____

_____

Step 5: _____

_____

Step 6: _____

_____

Step 7: _____

_____

Step 8: _____

_____

Step 9: _____

_____

Step 10: _____

_____

2. After trying the ten steps to better test taking, develop your own personalized test-taking steps.

Step 1: _____

Step 2: _____

Step 3: _____

Step 4: _____

Step 5: _____

Step 6: _____

Step 7: _____

Step 8: _____

Step 9: _____

Step 10: _____

Additional steps: _____

_____

_____

_____

_____

_____

7.3 Six Types of Test-Taking Errors

To improve future test scores, you must conduct a test analysis of previous tests. In analyzing your tests, you should look for the following kinds of errors:

1. Misread-directions errors
2. Careless errors
3. Concept errors
4. Application errors
5. Test-taking errors
6. Study errors

Students who conduct math test analyses will improve their total test scores.

Misread-directions errors occur when you skip directions or misunderstand directions but do the problem anyway.

EXAMPLES

Suppose you have this type of problem to solve:

$$(x + 1)(x + 1)$$

Some students will try to solve for x, but the problem calls only for multiplication. You would solve for x only if you have an equation such as $(x + 1)(x + 1) = 0$.

Another common mistake is not reading the directions before doing several word problems or statistical problems. All too often, when a test is returned, you find only three out of the five problems had to be completed. Even if you did get all five of them correct, it cost you valuable time that could have been used obtaining additional test points.

To avoid misread-directions errors, carefully read and interpret all the directions. Look for anything that is unusual, or notice if the directions have two parts. If you do not understand the directions, ask the instructor for clarification. If you feel uneasy about asking the instructor for interpretation of the question, remember the instructor in most cases does not want to test you on your interpretation of the question but how you answer it. Also, you don't want to make the mistake of assuming that the instructor will not interpret the question. Let the instructor make the decision to interpret the question, not you.

Careless errors are mistakes that you can catch automatically upon reviewing the test. Both good and poor math students make careless errors. Such errors can cost a student a higher letter grade on a test.

> ### EXAMPLES
>
> *Dropping the sign:* $-3(2x) = 6x$ instead of $-6x$, which is the correct answer.
>
> *Not simplifying your answer:* Leaving $(3x - 12)/3$ as your answer instead of simplifying it to $x - 4$.
>
> *Adding fractions:* $1/2 + 1/3 = 2/5$ instead of $5/6$, which is the correct answer.
>
> *Word problems:* $x = 15$ instead of "The student had 15 tickets."

However, many students want all their errors to be careless errors. This means that the students did know the math but simply made silly mistakes. In such cases, I ask the student to solve the problem immediately, while I watch.

If the student can solve the problem or point out his or her mistake in a few seconds, it is a careless error. If the student cannot find an error immediately, it is not a careless error and is probably a concept error.

When working with students who make careless errors, I ask them two questions. First, I ask, "How many points did you lose due to careless errors?" Then I follow with, "How much time was left in the class period when you handed in your test?" Students who lose test points due to careless errors are giving away points if they hand in their test papers before the test period ends.

To reduce careless errors, you must realize the types of careless errors that students make and

recognize them when reviewing your test. If you cannot solve a missed problem immediately, it is not a careless error. If your major error is not simplifying answers, review each answer as if it were a new problem and try to reduce it.

Concept errors are mistakes made when you do not understand the properties or principles required to work math problems. Concept errors, if not corrected, will follow you from test to test, causing a loss of test points.

> ### EXAMPLES
>
> Some common concept errors are not knowing
> - $(-)(-)x = x$, *not* $-x$
> - $1(2) > x(-1)$ gives $2 < x$, *not* $2 > x$
> - $5/0$ is undefined, *not* 0
> - $(a + x)/x$ is *not* reduced to a
> - the order of operations

Concept errors must be corrected to improve your next math test score. Students who make numerous concept errors will fail the next test and the course if concepts are not understood. Just going back to rework a concept error problem is not good enough. You must go back to your textbook or notes and learn why you missed that type of problem, not just the one problem itself.

The best way to learn how to work the types of problems on which you tend to make concept errors is to set up a concept error problem page in the back of your notebook. Label the first page "Test One Concept Errors." Write down all your concept errors and how to solve the problems. Then, work five more problems that involve the same concept. Now, in your own words, write the concepts that you are using to solve these problems.

If you cannot write a concept in your own words, you do not understand it. Get assistance from your instructor if you need help finding similar problems using the same concept or if you cannot understand the concept. Do this for every test.

Application errors occur when you know a concept but cannot apply it to a problem. Application errors usually are found in word problems, deducing formulas (such as the quadratic formula), and graphing. Even some better students become frustrated with application errors; they understand the material but cannot apply it to problems.

To reduce application errors, you must predict the types of application problems that will be on the test. You must then think through and practice solving those types of problems using the appropriate concepts.

EXAMPLE

If you must derive the quadratic formula, you should practice doing it backward and forward while telling yourself the concept used to move from one step to the next.

Application errors are common with word problems. After completing a word problem, reread the question to make sure you have properly applied the answer to the intended question. Application errors can be avoided by appropriate practice and insight.

Test-taking errors apply to the specific way you take tests. Some students consistently make the same types of test-taking errors. Through recognition, these bad test-taking habits can be replaced by good test-taking habits. The result will be higher test scores. The list that follows includes the test-taking errors that can cause you to lose many points on an exam:

1. Missing more questions in the first third, second third, or last third of a test is considered a test-taking error.

   Missing more questions in the first third of a test can be due to carelessness when doing easy problems or due to test anxiety.

   Missing questions in the last third of the test can result from the fact that the last problems are more difficult than the earlier questions or from increasing your test speed to finish the test.

   If you consistently miss more questions in a certain part of the test, use your remaining test time to review that section of the test first. This means you may review the last part of your test first.

2. *Not completing a problem to its last step* is another test-taking error. If you have this bad habit, review the last step of each test problem first, before doing an in-depth test review.

3. *Changing test answers from correct ones to incorrect ones* is a problem for some students. Find out if you are a good or bad answer changer by comparing the numbers of answers you change to correct and to incorrect answers. If you are a bad answer changer, write on your test, "Don't change answers." Change answers only if you can prove to yourself or the instructor that the changed answer is correct.

4. *Getting stuck on one problem and spending too much time on it* is another test-taking error. You need to set a time limit on each problem before moving to the next problem. Working too long on a problem without success will increase your test anxiety and waste valuable time that could be used in solving other problems or in reviewing your test.

5. *Rushing through the easiest part of the test and making careless errors* is a common test-taking error for the better student. If you have the bad habit of getting more points taken off for the easy problems than for the hard problems, review the easy problems first, and the hard problems later.

6. *Miscopying an answer from your scratch paper to the test* is an uncommon test-taking error, but it does cost some students points. To avoid this kind of error, systematically compare your last problem step on scratch paper with the answer written on the test. In addition, always hand in your scratch paper with your test.

7. *Leaving answers blank* will get you zero points. If you look at a problem and cannot figure out how to solve it, do not leave it blank. Write down some information about the problem, rewrite the problem, or try to do at least the first step.

8. *Solving only the first step of a two-step problem* causes some students to lose points. These students get so excited when answering the first step of the problem that they forget about the second step. This is especially true on two-step word problems. To correct this test-taking error, write "two" in the margin of the problem. That will remind you that there are two steps or two answers to this problem.

9. *Not understanding all the functions of your calculator* can cause major testing problems. Some students only barely learn how to use the critical calculator functions. They then forget or have to relearn how to use their calculators, which costs test points and test time. Do not wait to learn how to use your calculator during the test. Overlearn the use of your calculator before the test.

10. *Leaving the test early without checking all your answers* is a costly habit. Do not worry about the first person who finishes the test and leaves. Many students start to get nervous when students start to leave after finishing the test. This can lead to test anxiety, mental blocks, and loss of recall.

According to research, the first students finishing the test do not always get the best grades. It sometimes is the exact opposite. Ignore the exiting students, and always use the full time allowed.

Make sure you follow the ten steps to better test taking. Review your test-taking procedures for discrepancies in following the ten steps. Deviating from these proven ten steps will cost you points.

Study errors, the last type of mistake to look for in test analysis, occur when you study the wrong type of material or do not spend enough time on pertinent material. Review your test to find out if you missed problems because you did not practice those types of problems or because you did practice them but forgot how to do them during the test. Study errors will take some time to track down, but correcting study errors will help you on future tests.

If you are like most students, after analyzing one or several tests, you will recognize at least one major, common test-taking error. Understanding the effects of this test-taking error should change your study techniques or test-taking strategy.

EXAMPLE

If there are 7 minutes left in the test, should you review for careless errors or try to answer those two problems you could not totally solve? This is a trick question. The real question is, do you miss more points due to careless errors or concept errors, or are the missed points about even? The answer to this question should determine how you will spend the final minutes of the test. If you miss more points due to careless errors or miss about the same number of points due to careless errors and concept errors, review for careless errors.

Careless errors are easier to correct than concept errors. However, if you make very few or no careless errors, you should be working on those last two problems to get the greatest number of test points. Knowing your test-taking errors can add more points to your test by changing your test-taking procedure.

Now that the Six Types of Test-Taking Errors have been reviewed a real life example may be helpful in applying this analysis to tests. Question four in the Section Review pertains to reviewing Figure 17 and recording the error types, points lost for each error, and examples of errors. Since this is a real test not all the test-taking errors are used and some errors are used twice. The answer key is at the end of the Section Review.

FIGURE 17 Math Test for Prealgebra

The answers are in boldface. The correct answers to missed questions are shaded. Identify the type of error based on the Six Types of Test-Taking Errors. The student's test score is 70.

1. Write in words: 32.685

   Thirty-two and six hundred eighty-five thousandths

2. Write as a fraction and simplify: 0.078

   78/1000   **39/500**

3. Round to the nearest hundredth: 64.8653

   64.865   **64.87**

4. Combine like terms: 6.78x – 3.21 + 7.23x – 6.19

   = 6.78x + 7.23x + (–3.21) + (–6.19)

   = 14.01x – 9.4

5. Divide and round to the nearest hundredth: 68.1357 ÷ 2.1

   32.4454 → 32.45

6. Write as a decimal: $\frac{5}{16}$

   0.3125

7. Insert < or > to make a true statement:

   $\frac{3}{8}$ < $\frac{6}{13}$

8. Solve: $\frac{3}{x} = \frac{9}{12}$

   $9x = 3(12)$

   $\frac{x}{9} = \frac{36}{9}$

   $x = 5$ **x = 4**

9. What number is 35% of 60?

   2100   **21.00**

10. 20.8 is 40% of what number?

    52

11. 567 is what percent of 756?

    $\frac{756}{567} = \frac{9}{100}$ **-8**

    = 133.3%   **75%**

12. Multiply: **-2**

    (–6.03) (–2.31) = 13.9   **13.9293**

Answer Key for Prealgebra Test

1. Correct

2. *Misread-directions error*—Forgot to simplify by reducing the fraction.

3. *Concept error*—Did not know that hundredths is two places to the right of the decimal.

4. Correct

5. Correct

6. Correct

7. Correct

8. *Careless error*—Divided incorrectly in the last step.

9. *Test-taking error*—Did not follow Step 5 in test taking steps: reviewing answers to see if they make sense. The number that equals 35% of 60 can't be larger than 60.

10. *Test-taking error*—Did not follow Steps 7 and 8 in test-taking steps: don't leave an answer blank.

11. *Application error*—Solved the equation correctly but the equation setup was wrong.

12. *Concept error*—Did not know that when you multiply with one number in the hundredths, the answer must include the hundredths column.

SECTION 7.3 REVIEW

1. List the six types of test-taking errors.

   Type 1: _____

   Type 2: _____

   Type 3: _____

   Type 4: _____

   Type 5: _____

   Type 6: _____

2. Which of these types of errors have you made in the past?

   _____

   _____

3. How can you avoid these types of errors on the next test?

   _____

   _____

4. Review the Math Test for Prealgebra (Figure 17) if you are in a prealgebra course. For each error, list the type of error, the points lost for the error, and a different example of that type of error.

   Error Type: _____   Points Lost for Error: _____

   Example of Error: _____

   Error Type: _____   Points Lost for Error: _____

   Example of Error: _____

   Error Type: _____   Points Lost for Error: _____

   Example of Error: _____

   Error Type: _____   Points Lost for Error: _____

   Example of Error: _____

   Error Type: _____   Points Lost for Error: _____

   Example of Error: _____

   Error Type: _____   Points Lost for Error: _____

   Example of Error: _____

   Error Type: _____   Points Lost for Error: _____

   Example of Error: _____

5. Review your last test and list the types of errors you made, the points lost for each error, and examples
    of the errors. You may have more errors than allowed for in the list below. Use a separate sheet of paper
    if you need to list more. (Remember: careless errors are problems you can solve immediately.)

Error Type: _____   Points Lost for Error: _____

Problem with the Error in It: _____

Error Type: _____   Points Lost for Error: _____

Problem with the Error in It: _____

Error Type: _____   Points Lost for Error: _____

Problem with the Error in It: _____

Error Type: _____   Points Lost for Error: _____

Problem with the Error in It: _____

Using the pages in the back of your math notebook, rework the problems on which you made concept
errors and do five math problems just like each one. For each concept error, write why you could not
solve the problem on the test and what new information you learned that allowed you to solve the problem. Do this for every major test. Give an example of one of your concept errors and the reasons why
you can now solve the problem.

Concept Error: _____

Reasons Why I Can Now Solve the Problem: _____

_____

_____

7.4   How to Prepare for the Final Exam

The first day of class is when you start preparing for the final exam. Look at the syllabus or ask the instructor if the final exam is cumulative. A cumulative exam covers everything from the first chapter to the last chapter. Most math final exams are cumulative.

The second question you should ask is if the final exam is a departmental exam or if it is made up by your instructor. In most cases, departmental exams are more difficult and need a little different preparation. If you have a departmental final, you need to ask for last year's test and ask other students what their instructors say will be on the test.

The third question is how much the final exam will count. Does it carry the same weight as a regular test or, as in some cases, will it count as a third of your grade? If the latter is true, the final exam will usually make a letter grade difference in your final

grade. The final exam could also determine if you pass or fail the course. Knowing this information before the final exam will help you prepare for it.

Preparing for the final exam is similar to preparing for each chapter test. You must create a pretest to discover what you have forgotten. You can use questions from the textbook chapter tests or questions from your study group.

Review the concept errors that you recorded in the back of your notebook labeled "Test 1," "Test 2," and so on. Review your problem log for questions you consistently miss. Even review material that you knew for the first and second tests but that was not used on any other tests. Students forget how to work some of these problems.

If you do not have a concept-error page for each chapter and did not keep a problem log, you need to develop a pretest before taking the final exam. If you are preparing for the final individually, then copy each chapter test and do every fourth problem to see what errors you may make. However, it is better to have a study group where each of the four members brings in ten problems with the answers worked out on a separate page. Then each group member can take a 30-question test made up from the problems brought in by the other group members to find out what they need to study. You can refer to the answers to help you solve the problems that you miss. Remember, that you want to find out what you don't know before the test, not during the test. Developing and using these methods before each test will improve your grades.

Make sure to use the ten steps to better test taking and the information gained from your test analysis. Use all the time on the final exam, because you could improve your final grade by a full letter if you make an A on the final.

SECTION 7.4 REVIEW

What are the three questions you need to ask about your final exam?

Question 1: _____

Question 2: _____

Question 3: _____

7.5 Motivating Yourself to Learn Math

This text has taught you the skills necessary to improve your math learning and success. Now is the time to apply these new math study skills and test-taking behaviors and to eliminate the old behaviors that may have caused previous math failure. To accomplish this task you need to motivate yourself to use these skills, which can involve self-esteem, locus of control, and self-efficacy. No longer can you use poor math study skills and test-taking skills as an excuse for not being successful in math. Now is the time to take more responsibility for your success. Thousands of students have taken up this responsibility, passed math and graduated. In the process, they have also changed their lives.

First, let's take a look at self-esteem. Self-esteem is the part of the personality that allows us to feel good about ourselves. Many students have positive self-esteem in their personal lives but do not in regards to their abilities to succeed in math or college. Other students have positive self-esteem in college but not in their personal lives. Having positive self-esteem means that you can accept yourself and then develop a "can-do" attitude that is geared toward accomplishing your goals. Students with poor self-esteem may not put forth as much effort in accomplishing their goals. They give up easily. The first step in recovering from a low self-esteem is to replace negative thoughts with positive thoughts—thoughts that express a willingness to succeed in life instead of worrying about failure. This can be accomplished by setting short-term goals and taking pride in any subsequent successes.

The next area of taking responsibility involves the locus of control. A locus of control is the degree to which a student claims control over his or her own life. Do students feel that they are in control of their own lives, or is it their environment that dictates their choices and actions? Students who feel that they are not in control of their lives are considered "external" students, while students who feel they are in control of their lives are referred to as "internal." Internal students accept the responsibility of their behaviors and adjust their behaviors in order to become more successful. For example, the internal person does not blame instructors for his/her poor grades. This student also realizes that he or she can change his/her study behaviors instead of simply complaining about things that are out of his or her control. Are you internal or external?

Finally, there is self-efficacy. Self-efficacy is a new concept that is related to locus of control but is more specific to the belief that you can accomplish a specific goal. You need to ask yourself if you have the skills to be successful at _____. You can fill in the blank with tasks such as running a marathon, getting a better job, passing English or passing math. You may have self-efficacy in one or two of these areas but not all of them. Self-efficacy is an important success predictor as to how likely you are to succeed in math or any other of these other areas. In other words, students who believe they can be successful in math have a higher chance of passing a math course. With these new math study skills and test-taking skills you've developed through this workbook, your self-efficacy should have improved in the area of succeeding in math. This means that you should become more persistent and that if certain learning and test-taking behaviors do not work out, you should change them and try different techniques from this text. Having self-efficacy in math means that you will keep trying to succeed even if it means getting help from others.

Motivation is internal but you may need some help in getting it out. If you need extra help in improving motivation, talk to your instructor or contact your advisor or counselor to help you start the process. This does not mean that you have to like math, you just have to pass it. Now is the time to tell yourself that **I CAN BECOME SUCCESSFUL IN MATH.**

SECTION 7.5   REVIEW

1. Define self-esteem and explain how it affects math learning.

   _____

   _____

   _____

2. Define locus of control and explain how it affects math learning.

   _____

   _____

   _____

3. Define self-efficacy and how it affects math learning.

   _____

   _____

   _____

4. Who should you contact to help you improve your motivation to learn math?

   _____

   _____

   _____

CHAPTER 7   REVIEW

1. You should start studying for a test at least _____ days ahead of time.

2. You should be reviewing only the material you have _____ the night before the test.

3. When previewing a test, put a _____ by each question you can solve without thinking, and do those problems _____.

4. When taking a test, you must decide the best way to get the _____ points in the _____ amount of time.

5. Halfway through the test, you should have _____ than half of the problems completed.

6. If you cannot write down the first step of a problem, you should _____ the problem to help you remember how to solve it.

7. Do not leave the test room until you have reviewed each problem _____ times or until the bell rings.

8. In reviewing your test, if you cannot find an error immediately (within a few seconds), it is not a _____ error and is probably a _____ error.

9. Students who make numerous _____ errors will fail the next test and the course if the _____ are not understood.

10. The _____ day of class is when you start preparing for the final exam.

What is the most important information you learned from this chapter?

_____
_____
_____

How can you immediately use it?

_____
_____
_____

DAN'S TAKE REVIEW

1. Why did Dan feel the need to rush through the test?

_____
_____
_____

2. What often resulted from his fear of being the last to complete his exam?

_____

_____

_____

3. How did he eventually overcome this problem?

_____

_____

_____

INDEX

A

Abbreviations, 55
Abstraction anxiety, 33
Academic achievement, 14–17
Acronyms, 99
Active listening, 53
Affective characteristics, 14–16, 25–26
Algebraic symbols and expressions, 78
Algorithms, 102
Analytic learners, 22
Anxiety. *See also* Test anxiety
   causes of, 33–34
   defined, 32–33
   effect on learning, 35
   Exploring Past Math Experiences Worksheet, 34–35
   facts about, 37
   helpful suggestions, 35–36, 110–111
   measuring, 25
   types of, 33
Application errors, 117–118
Association learning, 99
Attendance, 108
Attitudes
   determining yours, 25
   negative, 9, 43–46
   positive, 98, 110
   self-efficacy, 124
   self-esteem, 123
Auditory learners, 93–94
Autobiography, 46

B

Bloom, Benjamin, 14

C

Calculator handbook, 61
Careless errors, 117
Classroom participation, 60
Classroom seating, 52
Cognitive anxiety, 40

Cognitive entry skills, 14–16
Cognitive learning styles, 17–23
Commonsense learners, 22
Computer/online resources, 81–82
Concept errors, 117
Conditioning, 88–89
Cue-controlled relaxation, 43

D

Data dump, 112–113
Deep breathing, 42–43
Dynamic learners, 22

E

Early morning classes, 53
Emotional anxiety, 40
Errors in test-taking, 116–117
Exams. *See* Tests
Exploring Past Math Experiences Worksheet, 34–35

G

Glossaries, 59
Goal setting, 71
Golden triangle of success, 52
Grades
   high school *vs.* college, 7
   passing, 108–109
   in previous courses, 15–16, 18, 26
Group learners, 95–96

H

High school *vs.* college math, 6–8
Homework. *See also* Notes and note taking; Review; Study skills
   falling behind, 75
   importance of, 72–73
   math anxiety and, 35, 40
   note cards, 77–78
   passing grades and, 108–109
   procrastination and, 40–41